Colours and Colour Vision

Colours are increasingly important in our daily life but how did colour vision evolve? How have colours been made, used and talked about in different cultures and tasks? How do various species of animals see colours? Which physical stimuli allow us to see colours and by which physiological mechanisms are they perceived? And how and why do people differ in their colour perceptions? In answering these questions and others, this book offers an unusually broad account of the complex phenomenon of colour and colour vision. The book's broad and accessible approach gives it wide appeal; it will serve as a useful coursebook for upper-level undergraduate students studying psychology, particularly cognitive neuroscience and visual perception courses, as well as for students studying colour vision as part of biology, medicine, art and architecture courses.

Daniel Kernell is Emeritus Professor of Medical Physiology at the University of Groningen, The Netherlands.

Colours and Colour Vision

An Introductory Survey

DANIEL KERNELL
University of Groningen, The Netherlands

CAMBRIDGE
UNIVERSITY PRESS

Shaftesbury Road, Cambridge CB2 8EA, United Kingdom

One Liberty Plaza, 20th Floor, New York, NY 10006, USA

477 Williamstown Road, Port Melbourne, VIC 3207, Australia

314–321, 3rd Floor, Plot 3, Splendor Forum, Jasola District Centre, New Delhi – 110025, India

103 Penang Road, #05–06/07, Visioncrest Commercial, Singapore 238467

Cambridge University Press is part of Cambridge University Press & Assessment,
a department of the University of Cambridge.

We share the University's mission to contribute to society through the pursuit of
education, learning and research at the highest international levels of excellence.

www.cambridge.org
Information on this title: www.cambridge.org/9781107083035

© Daniel Kernell 2016

First published 2016

A catalogue record for this publication is available from the British Library

Library of Congress Cataloging-in-Publication data
Kernell, Daniel.
Colours and colour vision : an introductory survey / Daniel Kernell, University of Groningen, The Netherlands.
    pages   cm
Includes bibliographical references and index.
ISBN 978-1-107-08303-5 (Hardback : alk. paper) – ISBN 978-1-107-44354-9 (Paperback : alk. paper)
1. Color vision.   2. Color blindness.   3. Visual perception.   4. Optics.
I. Title.   II. Title: Colors and color vision.
QP483.K47 2015
612.804–dc23   2015018984

ISBN   978-1-107-08303-5   Hardback
ISBN   978-1-107-44354-9   Paperback

To Hilda

Contents

Colour plate section between pages 138 and 139.

Plates

Figures

Tables

Preface

This is a book about many different aspects of colours, how they arise and how one might see and experience them. When writing this book, my first source of inspiration was my own visual system: I belong to the rather large minority with an inherited red-green blindness. It has often astonished me that most people know so little about what this sensory constitution means, in spite of the fact that, in our part of the world, it affects more than 4% of the total population. Thus, I started my writing enterprise as a book about colour blindness, but the project gradually expanded to become a more general survey of matters concerning colour. The description includes an account of the physical and physiological mechanisms of colours and colour vision in humans and other animals, which comes naturally to me because I worked in neurophysiological research for many years (albeit on subjects other than colour vision).

Colours often give us a very direct and immediate kind of sensory experience and one might therefore be inclined to think that the nature of the phenomenon is simple and straightforward. This is, however, not the case: colour vision is a highly complicated and multidimensional subject matter. For many people, colours are an important source of enjoyment in everyday life, in nature and in various expressions of art and culture (true also for red-green colour-blind persons). Publications about colour often mainly deal with their various aesthetic qualities. In 1819, Keats published his very long poem, *Lamia*, which includes a few famous lines suggesting that the rainbow might lose its colourful beauty if one knows too much about it:

> *Philosophy will clip an Angel's wings,*
> *Conquer all mysteries by rule and line,*
> *Empty the haunted air, and gnomed mine*
> *Unweave a rainbow*[1]

However, it might equally well be argued that the unweaving of a rainbow does not make its colours and beauty less impressive but rather the opposite: the more one knows about a subject the more interesting and captivating it usually becomes.[2] According to some interpretations of Keats' poem, the author himself and his contemporary colleagues might even have agreed on this point, provided that one does not lose one's sense of wonder when confronted with the many complexities of human perceptions and the natural world.

Administrative note

Supplementary information concerning items in the running text may be found in notes at the end of the book. In Appendix E, an explanatory list is given for commonly used technical terms. In order to facilitate reading of the book, greytone-versions of all illustrations are shown close to the relevant sections of running text. For illustrations referred to as 'Plates', which provide further information by using coloured components, additional full-colour prints are included in a separate colour plate section.

The author of this book has lived in Sweden, in the Netherlands and (briefly) in Britain. When mentioning 'our culture' or other similar concepts, the 'our' or 'us' or 'we' refers to inhabitants or conventions of Western Europe.

1   Colour vision in everyday life

In my childhood, the surrounding world was rather sparsely coloured. In the 1930s to 1950s, photographs and movies were typically in black and white and there were no TVs or computers and, hence, no associated colour monitors. Typewriters wrote in black (sometimes also in red), and important sections of a text were not highlighted but simply underlined with a pencil. The telephone was immobile and typically black. Books about the visual arts were largely illustrated in greyish halftones, perhaps supplemented with a small number of expensive colour plates. Since then, colours have exploded into almost all sections of modern life, at work as well as in leisure activities: colour TV, computers with colour monitors and colour printers, digital cameras with colour sensors, even telephones take colour snapshots, and magazines and newspapers are illustrated in full colour. This present-day ubiquity of colour probably does not signal a sudden change in human culture and preferences but rather it reflects the power of modern electronic and chemical technology, allowing the innate human fascination with colour to become more fully expressed.

In this introductory chapter, I write a little about the practical importance and use of colours, their names, history and cultural significance, things one might discuss without much knowledge about the colour mechanisms. For instance, how many different colours can one actually see? How can they be described and labelled?

1.1 Numbers and dimensions of colours: hue, colourfulness, brightness

Laptop manuals often mention the mind-boggling number of about 16 million colours that may be seen on the monitor screen.[3] However, this only means that, technically, the screen can produce about 16 million different light qualities. It does not tell you how many of these qualities the human eye can distinguish from each other, seeing them as different colours; many of the 16 million colour settings will actually look the same to a human observer. Still, the total number of distinguishable colours is very large. For the analysis of this complex landscape, it is practical to consider its three main dimensions (Plate 1.1):

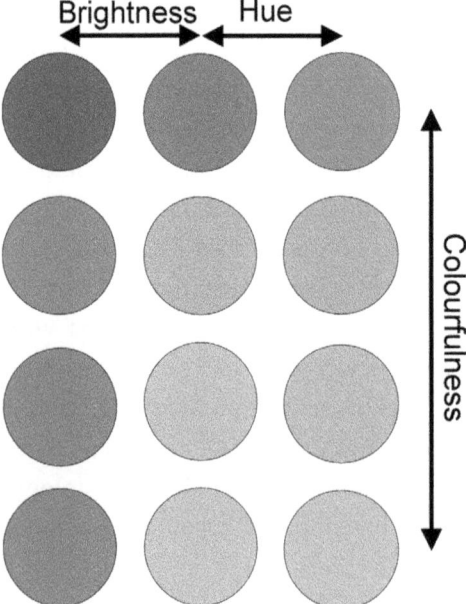

Plate 1.1 The three dimensions of colour: "*Brightness*" (approximate synonyms: *lightness / value*), here darker vs lighter blue. "*Hue*" (here blue vs. yellow). "*Colourfulness*" (approximate synonyms: *chroma / saturation*) here decreasing from above to below.
A black and white version of this figure will appear in some formats. For the colour version, please refer to the plate section.
In perception psychology, a distinction is often made between brightness versus lightness and between colourfulness versus chroma or saturation (cf. note 73; Hunt and Pointer (2011); in the present very general account, such technical distinctions will not be made between these various approximate synonyms. The RGB values used for the diagram: 0/0/153; 105/105/153; 135/135/153; 153/153/153 for the dark blue circles (upper to lower); 0/0/204; 140/140/204; 180/180/204; 204/204/204 for the lighter blue circles; 204/204/0; 204/204/140; 204/204/180; 204/204/204 for the yellow circles.

1. The *hue* of the colour, the property that we label with terms such as red, yellow, green, blue, etc.
2. The *colourfulness* of the colour (also often referred to as its saturation, purity or chroma), i.e. how strongly the hue is expressed. The colourfulness varies with admixtures of white/grey/black: the more white/grey/black one adds to a colour, the less saturated, pure and colourful it becomes (Plate 1.1, from upper to lower circles). Some colour names specifically concern unsaturated colours (e.g. rose, pink, brown).
3. The *brightness* of the colour (also often called lightness or value), i.e. the perceived intensity of the light reaching the eye. For colours of reflected light, the absolute level of brightness depends both on the intensity of illumination and on the reflectance of the material. The perceived relative brightness depends, of course, on the distribution of light in the surrounding scenery (for further comments on the complex issues of colour terminology, see note 73).

In direct comparisons between different coloured samples, a person with normal colour vision can distinguish between about 150 different hues, 100–150 levels of saturation and 100–200 levels of lightness. Taking all three colour dimensions into account, this implies that such a person can differentiate between at least $150 \times 100 \times 100 = 1.5$ million different colours; the maximal number might be about $150 \times 150 \times 200 = 4.5$ million. These numbers are very large, but they are still substantially lower than the 16 million colours mentioned in laptop ads. Besides, there are also a number of highly saturated colours, which our eyes are capable of seeing although computer screens cannot produce them (cf. Plate 2.6).

The number of isolated colours that can be named and reliably identified in stand-alone situations is much smaller than the huge multitude of colours that are distinguishable from each other in direct comparisons. This is, of course, largely due to the fact that colours are continuous variables; how many weights or lengths or temperatures would you be able to quantify directly without recourse to comparisons or measurements? A person with normal colour vision can directly identify an isolated colour as being red, green, yellow or orange, but has a very limited ability for pinpointing the colour with regard to its finer details of relative hue, saturation or lightness. Without direct comparisons, only about 15–30 different stand-alone colours can be safely identified and correctly named.[4] This agrees quite well with the limited number of colour names in everyday use (see Section 1.2). A much larger number of colour names does, of course, occur in relation to various vocations, professions and industries. However, in these contexts the colours are either externally named (e.g. labels on paint tubes) or they are identified using various sample collections and technical procedures (see Appendix B).

Our subjective experience of colour is strangely dualistic: besides the 'colourful' *chromatic* colours we also have a category of *achromatic* colours in the series white, grey, black, which may be defined as completely unsaturated colours, i.e. colours without a hue. Thus, the achromatic colours are, as it were, one dimensional: they vary only in lightness and not in hue or saturation (cf. bottom row in Plate 1.1). Consequently, they can easily be ordered on a linear scale from white via varying degrees of grey to black. In spite of their colourless nature, white, grey and black play an important role in all systems for the ordering and classification of colours (Appendix B). The paradoxical subdivision into chromatic and achromatic colours illustrates an important dualistic property of our visual system: in daylight our eyes and brain are capable of seeing colour but, simultaneously, they are also used for the detection and analysis of borderlines between lighter and darker regions in the surrounding world, i.e. for the processing of achromatic black-and-white contrast patterns (for further details and comments, see Chapter 4).

1.2 Commonly used colour names in different languages

Most of the colour terms used in everyday life concern the names of hues (red, orange, yellow, green, blue, violet, purple) and names of achromatic colours (white, grey, black). In addition, a few of the everyday colour names specifically refer to unsaturated colours (e.g. rose, pink, brown).

The British statesman William Gladstone (1809–98), besides his intense political activities, also conducted detailed studies of classical Greek literature,[5] and when reading Homer (written in the eighth century BC) was surprised to find that the major epic works, the *Iliad* and the *Odyssey*, contained few names of colour hues; Homer seemed mainly interested in achromatic colours. Gladstone's list of Homer's unambiguous colour terms includes white, black, yellow, red, violet and indigo (i.e. a kind of blue). In a later article, published in 1877,[6] Gladstone suggests that the scarcity of colour terms in Homer's work was perhaps caused by biological deficits in colour vision among the ancient Greeks. Gladstone suggested that perhaps the colour vision of later generations had become gradually 'trained' to discriminate between more hues and nuances, and that the results of this training had become heritable (which is not consistent with present-day scientific knowledge). Thus, in spite of the fact that much of Gladstone's very detailed analysis of Homer was a linguistic effort, he did not realize that the relative lack of Homeric colour terms might have more to do with language and culture than with the physiology of colour vision. Modern investigations have indeed shown that the extent and nature of the colour nomenclature may differ widely between different cultures (Plate 1.2).

Cultural differences in the classification of colours have mainly been investigated by anthropologists studying relatively isolated communities. A classical and very influential study of this kind was published in 1969 by Berlin and Kay. They performed a direct experimental study of 20 different languages (see below); in addition, they studied the colour terms in 78 other languages. They concentrated on the analysis of non-composite 'basal' colour terms, i.e. words which were in common everyday use within the target population.

In the experimental studies, the visiting anthropologists first performed introductory interviews with members of the population in order to make a list of all the colour terms in the local language. Then individual participants were shown a collection of 329 different colourcaps (from the Munsell system, see Appendix B), including black, white and seven grades of grey.[7] At first the participant was asked, for each local colour term, to select the cap which best represented the colour in question. Most participants found this choice of *focal* colours to be an easy task, and the same individual made very similar choices on different occasions. The next task concerned the range of validity for each colour term; this was generally considered to be much more difficult than the determination of focal colours. In this context, the participant was asked to select all caps that might belong to the same named colour category. Participants were then very hesitant and repeated measurements gave varying results for the same individual. This is, of course, in accordance with everyday experiences within our own culture: also Europeans with normal colour vision may often disagree with regard to how various 'intermediate' colours should be named, e.g. green or bluish-green, red or reddish-orange, etc.

With regard to the focus colours, the results of Berlin and Kay (1969) were very clear and surprisingly reproducible for comparisons between different languages (Table 1.1).[8] Whenever applicable (see below), the colour terms red, yellow, green and blue were represented by very similar test caps within different languages and cultures. Such a correspondence between languages is

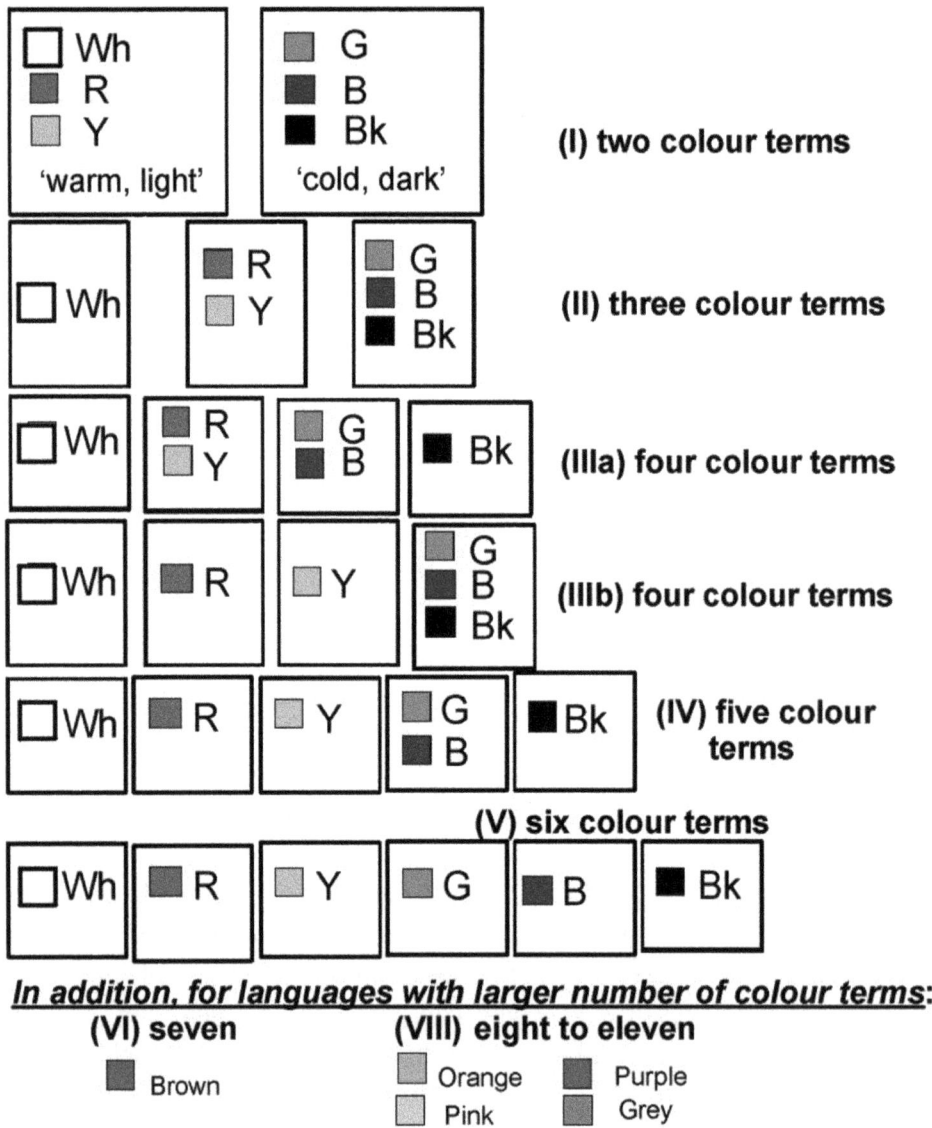

(I) two colour terms

(II) three colour terms

(IIIa) four colour terms

(IIIb) four colour terms

(IV) five colour terms

(V) six colour terms

In addition, for languages with larger number of colour terms:
(VI) seven   **(VIII) eight to eleven**
    ■ Brown   ☐ Orange   ■ Purple
                    ☐ Pink   ▨ Grey

Plate 1.2 Diagram summarizing the "validity-ranges" of colour names in languages with few colour terms. Small coloured squares signify each one of 11 basal colour names / categories in common European languages. The diagram shows in which manner these basal colours have generally been bundled together in languages with fewer terms. Depending on the number of basal colour terms, the languages were classified into seven different stages, here signified with Roman numbers I - VII. Note that languages of stages I-IV have one or several composite basal colour terms comprising >1 of our own most elementary colours of red, green, blue and yellow. *Abbreviations*: B blue, Bk black G green, R red, Wh white, Y yellow. A black and white version of this figure will appear in some formats. For the colour version, please refer to the plate section.
This illustration is based on data from Figure 17.13 in Kay and McDaniel (1997), and it gives a summary of findings from several investigators, including the classical study of Berlin and Kay (1969) and their successors. For a corresponding, and earlier, survey of 'focus colours', see Table 1.1.

Table 1.1 Focus colours in languages with different numbers of basal colour terms.

Number of colour terms	Focus colours
2	white, black
3	white, black, red
4	white, black, red, green *OR* white, black, red, yellow
5	white, black, red, green, yellow
6	white, black, red, green, yellow, blue
7	white, black, red, green, yellow, blue, brown
8, 9, 10, 11	*all above plus one or more of:* orange, purple, pink, grey

From results published by Berlin and Kay (1969). For a later and somewhat different analysis, see Figure 1.2.

not self-evident. Colour hues show a continuous variation (e.g. in a rainbow or solar spectrum) and, therefore, the general similarity of the four indicated focus colours across different cultures seems in accordance with the view that these four colours have some sort of special status in the physiology of human colour vision.[9]

In the most 'colour-poor' languages that still have general colour words, only two basal colour terms are used (e.g. Dani language, New Guinea; Plate 1.2, stage I). In modern Western languages about 11 basal colour terms are in everyday use (Plate 1.2, stage VIII). Between these extremes, focal colour terms are added in a remarkably similar way between different languages (Berlin and Kay, 1969): languages with only two terms have the focus colours 'white' and 'black' ('light' and 'dark'), languages with three terms have 'white' and 'black' plus 'red', etc. (Table 1.1).

There are no indications that people with different numbers of basal colour terms would differ in the number and quality of colours that they can actually see. Colour vision has been tested for people from many different cultures and countries, and only a minority of 4–5% or less have been found to have a permanent and inborn deviation of colour vision (Tables 5.2 and 6.1). In languages with few basal colour names, each term must represent a whole range of colour perceptions which are bundled together under a joint name, probably depending on how the colours were used within this linguistic region. In cultures with only two basal colour terms, colours apparently had a very limited relevance for the design of clothes, the performance of rituals, etc.

For the understanding of the role of colours in different cultures, the range of the various local colour terms is, of course, highly interesting. This problem was not extensively dealt with in Berlin and Kay (1969), but it has been analysed in several later studies by various investigators. As an example, I here show a summary of findings published by Kay and McDaniel (1997) (Plate 1.2). Also for such matters, variations between languages follow a general reproducible pattern. Comparisons between Table 1.1 and Plate 1.2 show that it is important to know both the focus and range aspects of colour linguistics: they are not directly predictable one from the other. After white and black, red is usually the first chromatic focus colour with a name of its own (Table 1.1); in stage-II languages this term does, however, represent a greater spectral region than the one we would call 'red'. Such composite colour terms may even have several, almost equivalent focal colour terms, e.g. red and yellow for the 'warm' composite colour in stage II (Plate 1.2, II). A colour term that has both a focus and a range corresponding to the European concept of 'red' appears only in languages with at least four basal colour terms (Plate 1.2, IIIb).

The linguistic study of colour classification is rapidly developing and some of the recently published models are mathematically and geometrically highly complex.[10] Following the classic study of Berlin and Kay (1969), a huge number of observations have been collected, and part of this material is even freely available for everyone via the internet.[11]

In the cultural evolution of colour names, green and blue have long been associated, sometimes collectively referred to with the English term 'grue'. This is, for instance, true for most of the indigenous American languages. Also among modern high-technology languages, the subdivision into different basal colours may sometimes differ. Russian has, for instance, no general name for 'blue' but makes a distinction between dark blue and light blue. However, it is obvious that the four colours blue, green, yellow and red appear rather early in the 'evolution' of colour terms, earlier than unsaturated chromatic colours (brown, pink) and earlier than colours that may be described as hue mixtures (orange = red + yellow; purple = red + blue).

Almost no language seems to have more than about 11 basal colour names in everyday use (Plate 1.2, VIII), i.e. colour categories that practically everybody who speaks this language uses (including colour-blind individuals, in spite of their risk of confusion). Of course, this does not mean that there are no more than 11 words for colours in such a language. Some of the common colour names are more or less synonymous with one of the basal terms (e.g. purple versus violet), and others are 'extra' everyday terms used for colours of specific objects (e.g. 'blond' for hair). Furthermore, as has already been mentioned, there are large numbers of colour names specifically used by various categories of professionals and specialists, e.g. artists, house painters, home decorators, etc. This is not a new phenomenon; for instance, hundreds of years ago very specific colour names were already used by horse traders to describe their animals: about 13 terms in the Latin of the fifth to seventh centuries, about 14 in thirteenth-century Spain and about 30 among present-day Kyrgyz.[12]

As one might expect, colour names are often derived from natural objects illustrating the meaning of the term. 'Green' is related to 'grow' and associated with the green of plants. 'Brown' is thought to be related either to 'bear' or to the German word for 'burned' (gebrannt). 'Rose' is, of course, named after the flower.

Sometimes colour terms may be difficult to interpret because they might be associated not only with the colour of a material but also with (other aspects of) the material itself. Thus, from about AD 900 until the Renaissance, the word 'purpura' was often used for a particular quality of silk fabric which could possess almost any colour, e.g. white, black, red, yellow, green, blue.[13] Only from the seventeenth century onwards did this term become used to indicate only one specific colour. A similar ambivalence existed for the word 'scarlet', a term originally used for an expensive kind of imported cloth, later on acquiring the associated meaning that the cloth was intensely red. After the import of cochineal from America started in the 1580s (Plate 1.4a), 'scarlet' started to become used as a stand-alone colour name, referring to crimson colours like those derived from the cochineal beetles.

1.3 Development of colour vision and colour naming in children

The physiological ability to distinguish different colours comes about very early during the development of children. Using various non-verbal methods, it has been found that a high level of colour discrimination is already reached in babies at 2 months of age.[14] Children start saying various basal colour words at the age of 2 years, but the appropriate use of these terms develops very gradually during the course of several years. A 3- or 4-year old might have a completely normal colour vision and still give you a wrong answer to the question, 'What colour is this?' Colour is an abstract concept and for children it is much easier and quicker to learn the names of common objects than to associate the correct colour term with the correct colour. The complexity of this process is suggested by the varying manners in which different colours are linguistically bundled together in different languages (Plate 1.2). During development, children usually learn to couple colour names to the correct colour most easily/quickly for red and then, in an order of increasing difficulty, for green, black, white, orange, yellow, blue, pink, brown and purple.[15] In this list, the six 'primary' or 'elementary' colours of red, green, black, white, yellow and blue precede most of the 'secondary' colours, i.e. colours that conveniently may be described as mixtures or transitions between the elementary ones. The only exception to this rule is the colour orange (= red + yellow), which might have had the advantage that in some languages (including English) orange is also the name of a fruit with the same colour.

The childhood development of correct colour naming takes place in roughly the same order as the evolution of focal colour names when comparing languages: also, in this latter case, the terms for primary colours precede those for the secondary ones, and red tends to be the first focal colour with a name of its own (Table 1.1, Plate 1.2). Results from a recent Dutch investigation give a concrete example of how young children deal with colour categories.[16] Colour samples were shown to the children and they were asked: 'what colour is this?' For the four chromatic elementary colours (red, green, yellow, blue) the 3-year olds gave a correct answer in 67% of the cases, and the corresponding numbers were 75% for the 4-year and 96% for the 5-year olds. In case of the secondary colours pink, orange and brown the scores for correct answers were clearly lower, particularly for the younger children. Average percentages of correct answers were then only 28% for the 3-year, 54% for the 4-year and 95% for the 5-year olds.[17] Only at the age of 6–7 years did the answers for all kinds of colours become 98–99% correct. Among the seven tested colours, orange and brown gave the lowest scores for the 3- and 4-year olds; in modern Dutch the words for the orange colour ('oranje') and the fruit ('sinaasappel') are quite different. The fact that children learn correct naming most quickly for the four elementary colours (red, green, yellow, blue) might partly, of course, depend on the frequency with which such colour terms are used by surrounding grown-ups and older children.[18]

In English, as well as in several other languages (e.g. Swedish and Dutch), the colour term typically precedes the name of the coloured object (e.g. 'the blue bird'), and this apparently makes things more difficult for the young child. In the course of a spoken sentence the colour is then, as it were, detached from the real world until the coloured object is mentioned.

Interestingly, experiments with children have shown that a correct association of colours with colour names is learnt more quickly if the word for the object precedes the name of the colour, like in French where 'the blue bird' becomes 'l'oiseau bleu'.[19]

1.4 History of colour pigments and dyes

Since ancient times, people have spent a lot of time and effort trying to discover ever more strikingly coloured stains and pigments. During human prehistory, colour pigments are included in the earliest items for long-distance commerce, in addition to materials for weapons and tools (e.g. flint, obsidian). Dyes and pigments have, since early times, been extracted from many different sources: minerals, plants and animals (Table 1.2).

Prior to the industrial revolution of colour manufacturing in the latter half of the nineteenth century, the painting artist's range of possible motifs was limited by the scarcity or lack of suitable pigments and paints, particularly concerning those for highly saturated hues. In ancient Greece, Aristotle pointed out that, due to a lack of suitable pigments, one could not depict the rainbow. This difficulty remained for a long time; in a lecture in 1669 at the French Academy, French artists were told that they should refrain from trying to paint brightly sunlit landscapes because no suitable pigments were available.[20]

The oldest colour pigments used for painting were earth pigments, often yellow, brown or reddish depending on their content of various iron compounds (e.g. red iron oxide, yellow iron hydroxide) or of compounds containing manganese (e.g. brown manganese dioxide). Earth pigments had already been used in Stone Age cave paintings. Colours of such pigments may be made darker and redder by heating, a method used for getting red ochre. Ochre pigments are widely available, non-poisonous and very stable. However, their colours are rather unsaturated; other substances were needed to get clear and strong colours. Compounds with such properties are also present among naturally occurring substances, but they are often difficult to find and may require complex processing methods.

The strong psychological impact of colours is illustrated by the large scale and dramatic circumstances of the early international trade in coloured pigments. The very exclusive purple stain was originally laboriously produced from secretions of molluscs living in eastern parts of the Mediterranean Sea (Plate 1.4b); roughly 8 500 000 snails were needed per kilogramme of dye. In ancient times,

Table 1.2 Original sources of some important pigments and dyes used before the 1850s.

Colour	Pigment/dye	Original source
Mineral origin[a]		
red	red ochre	earth pigment
	vermilion	mercury mineral (also early synthetic colour)
	Falu red	originally minerals from copper mines
orange	minium, red lead	mineral with lead tetraoxide
brown	umber	earth pigment with iron and manganese oxide
green	green earth	clay coloured by various metal compounds
	malachite	copper mineral
	chrome green (viridian)	chromium mineral
blue	ultramarine	mineral lapis lazuli
black	black earth pigment	manganese oxide
Plant origin[a]		
red	madder	roots of plant *Rubia tinctorum*
	brazilin	wood, e.g. from Pau brasil tree
yellow	saffron	stigmas of plant *Crocus sativus*[b]
	weld	leaves of plant *Reseda luteola*
blue	indigo	fermented leaves of e.g. *Indigofera tinctoria*
	woad	leaves of plant *Isatis tinctoria*
black	carbon black	wood charcoal
Animal origin[a]		
red	crimson, kermes	insect *Kermes vermilio*
	carmine	insect *Dactylopius coccus*, living on cactus
yellow	Indian yellow	possibly (?) from cow urine[c]
brown	sepia	inksac of cuttlefish (*Sepia*)
purple	purple dye	sea snail *Murex brandaris*[d]
black	bone black	charred bones
Synthetic		
yellow	cadmium yellow (1817)	cadmium sulfide
green	emerald (Schweinfurt) green (1814)	copper aceto-arsenate

Table 1.2 (*cont.*)

Colour	Pigment/dye	Original source
blue	Egyptian blue, caeruleum (BC)	calcium copper silicate
	cobalt blue (pure form, 1802)	cobalt oxide – aluminium oxide
white	white lead (= lead white) (BC)	lead carbonate and hydroxide
	zinc white (<1800)	zinc oxide

[a] For the mineral, plant and animal categories, the list shows only the original starting material used for the fabrication of various colours. Obtaining the actual dye/pigment often involved a more or less complex processing, sometimes obscuring the difference between e.g. the 'synthetic' and other source categories. For the synthetic category, the approximate year in which the stain first became available is shown.

[b] About >110 000 crocus flowers required per kilo of saffron; hence, a very expensive dye.

[c] *Indian yellow*, magnesium euxanthate, was commonly used by European artists before the end of the eighteenth century. Imported from India and China, it was originally claimed to have been extracted from the urine of Indian cows fed only on mango leaves and water. However, it is doubtful whether this story is true, it is not supported by much hard evidence (Finlay, 2003). For use in painting, the original pigment was later replaced by synthetic Indian yellow dye.

[d] Phoenicians often used the purple dye from *Murex brandaris* together with the pigment buccinum, obtained from the snail *Purpura haemastroma* (Gage 1993: p. 25).

purple dye was an important article in the international trade run by Phoenicians from Tyros and Sidon. The very expensive blue pigment ultramarine ('beyond the sea'), made from the mineral lapis lazuli, was partly derived from distant mines in mountains of Afghanistan (Plate 1.3a). Its high price caused the French government, in 1802, to ask the chemist Louis-Jacques Thénard to provide a cheaper replacement; this led to the further development of cobalt blue. Impure forms of this stain had already long been used in Chinese porcelain.

The urge to find strong and clear colouring stains for the textile industry provided one of the driving forces for many of the adventurous voyages of exploration to Africa and America in the fifteenth and sixteenth centuries. For many of the well-known classical coloured pigments very long-distance travel was indeed required, and this is partly reflected in their names. The country of Brazil got its own name from the Pau brasil tree (Plate 1.3b), the wood of which was ground to provide a red colour. The indigo colour and pigment got its name from India, its country of origin.

Some of the classical dyes of importance for textile staining were extracted from scale insects. Since very ancient times, the reddish crimson (kermes) dye had been obtained from the insect *Kermes vermilio*. After the Middle Ages this

Plate 1.3 (a) Colours from far away. **(a)** The blue pigment ultramarine, prepared by powdering the mineral *lapis lazuli*, was obtained from mines in e.g. Afghanistan.
Source: http://en.wikipedia.org/wiki/File:Natural_ultramarine_pigment.jpg.
Image by Palladian.

(b) The wood of the Pau Brasil tree (*Caesalpinia echinata*) delivered a red pigment; its home country got its name from this tree.
Source: http://en.wikipedia.org/wiki/File:PAUBRASILjbsp.jpg.
Image by Mauroguanandi.
A black and white version of this figure will appear in some formats. For the colour version, please refer to the plate section.

dye was largely replaced by carmine dye from cochineals, insects living on certain kinds of cactus (Plate 1.4a). Like the cactus plants themselves, the cochineals and the carmine dye originally came from America, and the dye was used already by the Aztecs. More than 80 000 female insects were needed to produce 1 kg of the red carmine dye. The early importance of this pigment is, for instance, illustrated by the agricultural history of the island Gran Canaria. In the middle of the nineteenth century much of its agricultural space was used for growing cactus plants with cochineals. Hence, in Gran Canaria, a major economic crisis was caused by the rise of the modern colour industry, with the marketing of cheap synthetic aniline colours beginning in the 1860s.

Besides finding them in nature, coloured dyes and pigments may also, of course, be fabricated using chemical/physical techniques. Occasionally, this had already happened very long ago. Egyptian blue seems to be the earliest known colour substance of this kind,[21] and white lead was produced using chemical methods already in ancient Greek and Roman times. From (pre-)medieval times until the eighteenth century, alchemists were busy trying to employ their 'magical' kinds of chemistry to produce silver and gold. Their studies and experiments included observations of colour shifts that took place during various physical/chemical processes. On a few occasions, their efforts acciden-tally led to useful results, not as silver or gold but in the shape of other newly emerging substances. One of the oldest European examples of a synthetic chromatic pigment is the scarlet-coloured *vermilion*, which fourteenth-century alchemists managed to fabricate from sulfur and mercury. The general signifi-cance of this invention was, however, somewhat limited. Vermilion had already been synthesized in China in the eighth century, and since even more ancient times the substance had been collected as a mineral from mines in, for example, Spain, India and China.

Some of the classical colour pigments were poisonous and potentially dan-gerous for the user. This was true for vermilion and, to a high degree, for the classical and much used white-lead stain. This poisonous compound is particu-larly dangerous because, unfortunately, it has a pleasantly sweetish taste. Only since the nineteenth century has the use of white lead been replaced by non-toxic alternatives, like zinc white and titan white. Another toxic lead-stain was the red or orange *minium*, known already in ancient Rome. Its name was derived from the Spanish Minius River. Medieval book illustrators used minium in their small-scale illuminations, and these artists were sometimes referred to as *min-iators* and their work as *miniatures*.[22]

Plate 1.4 **(a)** Colours requiring lots of work. **(a)** Cochineal insects (*Dactylopius coccus*), used for red carmine dye, are cultivated on cactus plants. © The Newberry Library. (inset) Cochineal insect.
Source: http://en.wikipedia.org/wiki/File:Cochineal_drawing.jpg.

(b) The snail *Murex brandaris*, used for purple dye, can be found in the Mediterranean sea as well as in the Indian Ocean and the South China Sea.
Source: http://en.wikipedia.org/wiki/File:Haustellum_brandaris_000.jpg.
Image by M. Violante.
A black and white version of this figure will appear in some formats. For the colour version, please refer to the plate section.

The modern industrial fabrication of synthetic dyes started in the 1860s, following the 1856 discovery by William Henry Perkin (1838–1907) that violet 'mauveine' (= *aniline purple*) could be synthesized from coal tar components. After this, knowledge in this field developed very rapidly, and it soon became evident that similar techniques could be used to produce an enormous number of different colour dyes. Many of these substances were referred to as *aniline dyes* because they had been synthesized starting off with the colourless compound *aniline*, which itself was extracted from coal tar. Nowadays, one can buy dyes and pigments of almost any colour easily and cheaply. Before the middle of the nineteenth century, this was certainly not the case.

1.5 Psychology and symbolic values of colours

Chromatic colours may have marked psychological effects on people, which is illustrated by the symbolic use of colour terminologies in everyday language. A person with no particular characteristics is referred to as 'grey' and a flamboyant individual is 'colourful'. An emotionally charged and not altogether objective account may be 'coloured', and complex differences between various places might be described as variations in 'local colour' (French: *couleur locale*; Swedish: *lokalfärg*).

Colours often rapidly attract our attention, which might be one of their major biological functions, in addition to their use for the delineation and recognition of objects. Our perceptions of sound, taste and smell may differ in their emotional impact, some being experienced as pleasant and others as repellent. For many people, this is also true for various colours and colour combinations. As colours have an important role in several branches of aesthetic endeavour (e.g. pictorial art, textile industry, home decoration, etc.), it would be of great interest to know whether there are any general psychological principles or rules determining which colours or colour combinations are experienced as 'nice' and 'harmonious' and which seem 'awful' and 'distressing'. In this context, it is essential to remember that colours are three-dimensional entities: not only differences in hue but also variations in saturation and lightness may be of importance.[23] Several research teams have found that the more saturated the colours are, the more pleasant and attractive they become. The emotional reaction to the level of saturation seems to be stronger and more consistent than the reactions to differences in hue. With regard to hue variations, the most consistent result has been that, on average, subjects have a preference for blue.

In addition, there is commonly the opposite kind of response to yellow (i.e. a relative dislike). This latter reaction might, however, have been influenced by the fact that, for people with normal colour vision, yellow is an unusually unsaturated colour (cf. Figure 5.4). These preferences and dislikes are, of course, statistical truths, which means that they are not valid for all individuals. For instance, in one study, 56% of the participants preferred blue, 17% had an equal preference for blue and yellow, and 20% preferred either red or yellow.[24] In the evolution of colour-language complexity, the first chromatic focal colour to appear is not blue but red (Table 1.1), i.e. not the generally most pleasant of the hues.

Interestingly, a statistical preference for blue has also been reported for rhesus monkeys and pigeons,[25] and some inherited biological mechanism might conceivably exist that predisposes humans to like blue. The possible biological advantages of such a preference are unclear, and (additional) cultural influences are difficult to exclude. For instance, in China and Japan, blue is apparently less often a favourite hue than in countries with a European population.

There are also a few other instances of psychological colour effects which are possibly associated with inborn reaction patterns (although this has not yet been fully proven). In some such cases, the colour effects were clearly context bound, e.g. red colour having a signal function under particular conditions. Thus, in many non-human primate species, red colouring in females (e.g. of skin areas) increases their sexual attractiveness to males. In such cases, the red colouring might signal a fertility-related hormonal state of the female. Also in experiments on humans, females with associated red colouring were judged by males to be more attractive than those with other associated colours. This was, for instance, found for red colours in women's dress, and the effects were also present for red colours used in the background of photographs of the test women.[26]

Different kinds of context-associated effects of the colour red were seen in other recent series of experiments. During the 2004 Olympic Games, it was noticed that among participants in combat sports (e.g. boxing, wrestling), those wearing red won more often than those wearing blue. This was further investigated in controlled experiments with a random distribution of red and blue jerseys among the combatants: those wearing red had then, prior to the combat, a maximum voluntary arm and leg force that was marginally (10.5%) but significantly greater than that for their blue colleagues.[27] Intriguingly, somewhat corresponding effects of the colour red were even found in virtual multi-player combat games in which the digital 'gladiators' had red or blue clothing. The players of this 'Team Deathmatch' were anonymous and did not know about the investigation. The analysis of 1347 matches showed that red won in 54.9% of the matches; this difference from 50% was small but, statistically, highly significant.[28] Why the colour red would promote a victory in real or virtual combat situations is not known; again, there is speculation that in these cases the red colour effects might occur due to inborn reaction patterns to a context-related signal. For instance, in a situation of aggressive competition, the red face colour of a raging male might help to intimidate his adversaries and promote his own relative social dominance.

By far, most of our emotional reactions to colours and their hues are apparently not inborn but acquired by learning and cultural experience. There are many popular ideas about the 'biological' effects of colours for which there is a remarkable lack of broad scientific support. Thus, for instance, it has not been consistently confirmed that red colours are exciting and alarming, or that green ones are calming and relaxing. For example, in a recently published investigation in which care was taken to distinguish between effects of hue, brightness and saturation, the findings were precisely the opposite: the most arousing hues were variants of green.[29] The lack of any generalized truths with regard to such colour questions and impacts is illustrated by the enormous variety of cultural associations for each separate hue.[30] In the Western world, a white colour is associated with feast and marriage, in Asia it is often the colour of funerals and sorrow. Red is, among other things, a symbol for love, violence, stop, revolution, errors and Christmas joy.

Colours can apparently easily be charged with practically any kind of emotional or intellectual content, and that makes them very useful as symbols with a great degree of presence and flexibility. In modern life, colours are becoming increasingly important for the encoding of various kinds of information (including traffic lights, map information, recognition labels, etc.). This might cause problems for individuals who have a different colour vision from that of the population majority (see Chapters 5–6). The great power of colours as symbols and information carriers is probably of major importance for their many-sided use in various religious ceremonies and in association with 'alternative' diagnostic and therapeutic methods. An internet search, for instance, easily produces a great number of fanciful statements concerning the use of colours for the treatment of various health issues. Such applications of colours have a long history and are not supported by any solid scientific evidence. If a patient is told what he/she might expect from a particular kind of treatment, whether colour related or not, the chance is often relatively great that the expected therapeutic effects might turn up thanks to so-called *placebo mechanisms*, i.e. therapeutic effects that are more dependent on the expectations of the patient than on specific characteristics of the treatment.[31] Expectations and the associated internal psycho-physiological mechanisms may have dramatic effects on many body functions (e.g. pain), and placebo effects are important components in many kinds of medical treatment.

In spite of the strong psychological impact of colours, we have a rather unreliable memory for colour details and we often recall them as being more saturated than they actually were.[32] Furthermore, we may easily be influenced by misleading suggestions concerning the colours of past experiences. Deficiencies of our colour memory seem to be illustrated by the fact that colours are apparently easily lost from recalled dreams. In addition, this might exemplify that our sense of vision is mainly concerned with the recognition of objects and events using lightness contrast; in this context, colour has a secondary role (see Chapter 4). With regard to dreams, if colours are indeed remembered, they occur with frequencies which are typically ranked such that white ~ black > red > blue ~ yellow ~ green > brown > grey > orange > pink (average data for large numbers of dreamers[33]). Interestingly, this pattern resembles that for the occurrence of colour terms in literature (Figure 1.1; cf. also linguistic data of Table 1.1 and Section 1.3).

In the interesting *Stroop test*, named after J. R. Stroop, participants are confronted with differences between colour names and colours seen. The test contains words for various colours, each one written with coloured letters in a hue differing from that expressed by the word itself, e.g. 'blue' written with yellow letters, 'red' with blue letters, etc. Reading these words aloud is quicker than the naming aloud of their respective letter colours. This discrepancy becomes larger in situations of mental fatigue and lack of concentration, mental states in which the Stroop test may be useful as one of the methods for quantification.

1.6 Colour mixtures

1.6.1 Subtractive and additive mixtures

It has been known since ancient times that the world's multitude of perceived colours may be imitated using mixtures of only a few starting pigments.

Mixtures of colours behave differently from mixtures of sounds and tones: in a musical chord one can still perceive the presence of the component tones, whereas in a mixture of pigments or coloured lights the resulting colour shows no traces of the starting hues. In this context, it is important to make a distinction between the actual mixing of coloured lights or pigments and the purely linguistic method for describing transitional hues, using a 'mental mixture terminology' (e.g. 'a somewhat bluish light green'). Most perceived colour hues may be produced by very many alternative mixtures, and how a colour looks does not tell you by which mixture it was produced (for further details, see Plate 2.11 and Sections 2.6 and C.4).

Various issues concerning colour mixtures will be extensively dealt with further on in this book (e.g. Sections 2.6 and B.1.2). In this introductory context, it will suffice to know that, from a practical point of view, there are two major methods (Plate 1.5): (1) *subtractive mixing* of coloured dyes and pigments (e.g. in paintings); (2) *additive mixing* of coloured lights (e.g. in computer and TV screens).[34] These two mixing technologies may give anti-intuitive differences in their results: e.g. in subtractive mixtures of paints, blue and yellow often give green, whereas the result in additive mixtures of lights might be white.[35]

In subtractive mixtures, each component pigment will block (absorb) the light of certain colours (wavelengths) and reflect and/or transmit others (cf. Plate 1.5a). In paint mixtures, the various pigments will cover each other and together they will then absorb the sum of all the light components absorbed by each one of the individual pigments, i.e. similar to the effect of superimposed coloured filters (Plate 1.5a). Only the light that is not absorbed by any of the pigments (or filters) will remain to be reflected and reach our eyes. The more pigments there are in the mixture, the more of the light becomes absorbed and the less light will remain to be seen, i.e. subtractive mixtures become darker by adding more components to the mixture. Ultimately, for mixtures of many different pigments, the end result might be a muddy brownish black (Plate 1.5a).

Additive mixtures of lights of different hues may, for instance, easily be produced by projecting lights of different colour onto the same white screen, partly or completely overlapping with each other (Plate 1.5b). The more light sources one adds to the mixture, the lighter will the summed result become. With an appropriate choice of starting colours, the end result might become white.[36]

For whichever mixing method one might use, the colour perceived by the eye does, of course, depend on the mixture of light rays ultimately reaching it. Hence, for measurements of how colour vision works, the results of direct

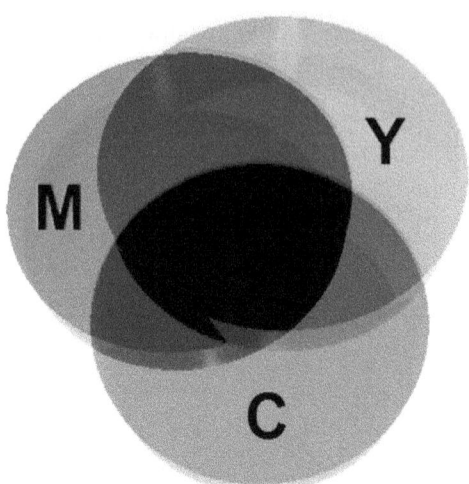

Plate 1.5 (a) Subtractive colour mixtures, using cyan (C), magenta (M), and yellow (Y) filters (cf. Plate 2.8).
Source: http://en.wikipedia.org/wiki/File:SubtractiveColor.svg.

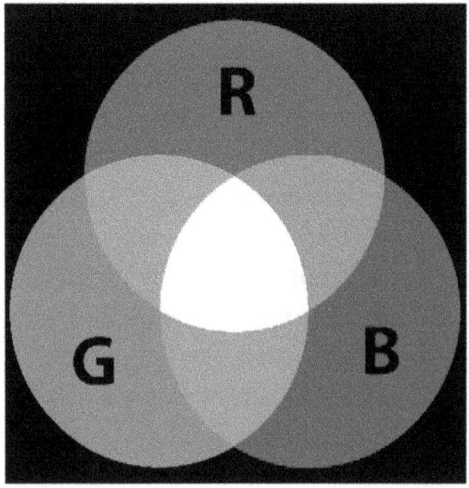

(b) Additive colour mixtures, using red (R), green (G) and blue (B) lights.
Source: http://en.wikipedia.org/wiki/File:AdditiveColor.svg.
A black and white version of this figure will appear in some formats. For the colour version, please refer to the plate section.

additive light mixtures are more directly and easily interpreted than the more complex effects of subtractive pigment mixtures. In pictorial art and in colours of everyday life, subtractive mixtures of pigments are, however, more common and more easily made.

1.6.2 Pigment mixtures and palettes

In painted pictures, the motif is usually rendered by subtractive mixtures of several different coloured pigments, covering each other. One of the advantages of, for instance, oil painting is that one might achieve the desired colour nuance by successively adding several thin layers of paint on top of each other, a clear example of subtractive colour mixture. The typical 'primary' colours for mixtures of paint are red, yellow and blue plus white and black; with these five starting colours practically all desired hues may be produced. Such mixing methods are, however, not ideal for producing highly saturated colours: the more kinds of different pigments one adds to the mixture, the less saturated the resulting final colour will become.

A painter's palette might be expected to contain at least the dyes for the five 'primary' hues, to be used for mixtures made during the course of the painting. Specific colour mixtures might be difficult to reproduce with sufficient precision, and for a painter it might therefore be convenient to include a number of pre-made mixtures on his palette. If certain strongly saturated colours are needed in addition to the primary ones, dyes for those extra hues might have to be added separately in order to avoid a degradation of saturation with pigment mixing. The Swedish artist K. G. Nilsson (2004) does, for instance, recommend at least 11 hues for the painter's palette: cadmium red, orange, light cadmium yellow, lemon yellow, green, blue green (turquoise), ultramarine blue, cobalt blue, violet, white and black.

In ancient times, artists apparently made little use of colour mixing while painting. The palette as an instrument for dye mixtures only started to be generally used in the fifteenth century, roughly from about the time when oil started to be used as a medium for the pigments. The palettes of known artists often contained about 10–20 different dyes, sometimes very much more. Delacroix (1798–1863) organized his starting colours in different ways depending on the project concerned. Sometimes, he provided his palette with more than 50 pre-mixed colour nuances, partly motivated by the fact that his large projects were executed together with assistants and Delacroix wanted to be certain that they all used his colour choices in a consistent manner.[37]

1.6.3 Optical mixtures

While subtractive colour mixtures have had a dominant role in most kinds of pictorial art, additive mixtures have also always been of significance. It has long been known that small spots or lines of different colour, placed close to each other, will merge to be seen as a joint mixed colour for a sufficiently distant observer. This kind of *optical additive mixture* is widely applied in modern imaging devices (e.g. colour televisions and computer monitors[38]), and it also concerns, for instance, the 'mixed' colours generated by multicoloured threads in cloth and by small stones in a mosaic (Plate 1.6). For classical mosaics, the principles underlying such optical colour mixtures were experimentally investigated and discussed by the Greek mathematician Klaudios Ptolemaios (~AD 90–168).[39] Additive colour mixtures also play an important role in textile art, e.g. in tapestries, consisting of thin multicoloured threads lying close to each other. These mixture effects were systematically studied and described in a notable book by the chemist Chevreul (1839), who was associated with the famous French tapestry factory Les Gobelins (Plate 1.6c). During the nineteenth century, the subject matter of optical colour mixtures was also dealt with in several other influential books. Such books were often directly aimed at the painting artist,[40] and techniques for promoting the appearance of additive colour mixtures in paintings were systematically and consciously applied by various artists, e.g. Signac and Seurat (pointillism; Plate 1.6b). Furthermore, in such paintings in which most of the hues are achieved by subtractive colour mixtures, additional coloristic effects due to optical mixtures probably often play a role as well for the viewer. The presence and importance of such effects would largely depend of the detailed colour structure of the painting.

1.7 Colour and pictorial art

The attractive powers of colours for humans are illustrated by their ceremonial use very early on during prehistory. The Neanderthal people, our prehistorical cousins, apparently did not make any figurative pieces of art, but they used red ochres for decoration at funerals. A similar use of red ochre was also common for Stone Age graves belonging to our own human kind (*Homo sapiens sapiens*); this red pigment is apparently the earliest example of a colour with a cultural importance among humans.

Plate 1.6 **(a)** Colour mixtures of mainly an additive (optical) kind take place in, for instance, mosaics **(a)**, paintings made using pointillistic techniques **(b)**, and woven fabrics **(c–d)**. **(a)** Roman mosaic from Tunisia.

(b) Example of pointillistic painting from a detail of *Circus Sideshow*, 1887–1888 by Georges-Pierre Seurat. © Tomas Abad/Alamy.

(c) Detail of tapestry of Louis XIV visiting the Gobelins Factory, 1673, by Charles Le Brun (1619–1690), 370 × 576 cm, Versailles, Musée National du Château. © DeAgostini Picture Library/Scala, Florence.

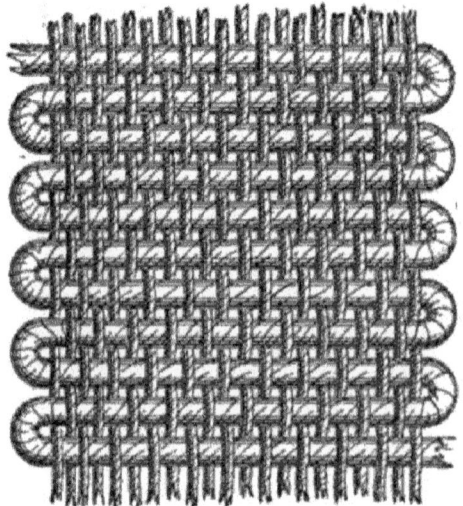

(d) General schematic illustration of woven fabric.
Source: http://en.wikipedia.org/wiki/File:Kette_und_Schu%C3%9F.jpg.
A black and white version of this figure will appear in some formats. For the colour version, please refer to the plate section.

The oldest examples of figurative art, cave paintings from the Stone Age in France and Spain, were initially largely monochromatic (outlines and major features in black coal or red ochre; e.g. Chauvet, about 30 000 years ago), but by about 25 000 years ago the first dichromatic paintings emerge, in black and red, e.g. in the French cave Pech Merle. At around 17 000 years ago, pictures were made with still greater numbers of colours, e.g. in the French caves Lascaux and Font-de-Gaume. Since then colours have continuously been of great importance in connection with the various manifestations of human creative art.

It is easily forgotten that the classical Greek statues were originally painted in vivid chromatic colours; this was rediscovered during the early half of the nineteenth century. However, for classical Greek artists, the relative lightness of the colours was felt to be more important than their relative saturation and hue variation;[41] often use was made of a rather limited number of different hues. Apelles (~356–308 BC) was one of the most famous painters of ancient Greece, and he was said to have used only four different hues in his art: black, white, red and yellow (it has been suggested that his 'black' included the hue 'blue'). The Roman scholar Pliny the elder (~AD 23–79) very much appreciated this simplicity and he deplored the fact that contemporary Roman painters used such a great number of unnecessary hues. Pliny died at the volcanic disaster of Pompeii. The paintings found in this Roman town were executed with pigments of about 29 different kinds,[42] i.e. a number considerably exceeding the Apellian four. However, not even Pliny had ever seen an Apelles painting, they were all gone by then and only their fame survived. For instance, we do not know to what extent Apelles and his colleagues produced variations in hue by mixing paint. In general, ancient and medieval painters seem to have had a certain reluctance to use pigment mixtures. Instead, nuances were sometimes created by placing strokes of different hues close to each other, i.e. an early method for optical (additive) colour mixture. The rich coloristic culture of the ancient Greek and Etruscan art is still the subject of much and difficult research, using modern techniques for reconstructing the original use of different colours. For instance, in marble statues of the sixth and early fifth century BC, all parts of the body and its clothes may have been covered with paint.

For Western paintings made in the period from the Middle Ages until the nineteenth century, colours were, of course, important, but their role was generally not a dominant one. Great emphasis was placed on structure and shape and from the fifteenth until the nineteenth centuries, reproductions of famous paintings were generally shown in black and white, often widely distributed as graphic prints.

In earlier centuries, the cost of pigments varied enormously, and this was an important factor determining their relative status and popularity. In the ancient world, purple colours were enormously expensive, and under the emperor Diocletian (~AD 245–311) purple-stained clothing was only allowed for members of the imperial staff; illegal use of this colour was severely punished. In Christian art, alterations in the price and status of coloured pigments influenced the hues of the robe of Virgin Mary. In Byzantine paintings, the robe was often stained purple. In the fifteenth-century art of the Low Countries it became scarlet red. In the Italian Renaissance, the robe was frequently ultramarine blue.[43]

Stained church windows represent particular and interesting aspects of medieval art. Light transmitted through stained glass may give strong colours with a much greater degree of saturation than those possible with light reflected from the pigments of a painting. Reflected colours will all be diluted by general reflections of the illuminating light; this kind of dilution does not happen for light transmitted through a stained-glass church window. In a medieval church, the 'magical' atmosphere is often further emphasized due to the darkening effects of a profusion of deep-blue glass. One of the first European Christian temples to be beautified with this kind of window was St Denis in Paris, in the twelfth century, under the enthusiastic guidance of Abbot Suger. At that time, blue-stained window glass was actually less expensive than the uncoloured variety. This was, however, not the main reason for the sacred bluish darkness; according to theologians of the time, God had his abode in darkness and, hence, the dark blue space of the church was an appropriate surrounding for religious meditation.[44] With regard to medieval symbols associated with blue, it is also interesting that one of the most valued gems of that time was the blue sapphire, which was thought to possess strong magical powers and which protected the carrier against many kinds of accident and disaster. In addition, the sapphire was believed to protect one's chastity and it was often carried by medieval priests.

In addition to their use in pictorial art, paints and pigments have, of course, been used for many other purposes, e.g. in buildings, furniture, mosaics, tapestries, clothing, make-up, etc. The colouring of textiles for clothing has long been of great economic importance, and in earlier times the status and price of the various pigments played a role similar to that described above for paintings: expensive colours were 'fine' colours. This has become radically altered in modern times: now the choice of colours is no longer dictated by their cost, practically all colours are available at a reasonable price. An earlier non-artistic use of colours for indicating social status is, for instance, illustrated by a Swedish example. A couple of hundred years ago, rich people started to paint their wooden houses red in order to make them resemble (at least from a distance) the more expensive houses that had been built using red bricks.[45] Later on, this practice became popular also for rural houses and farms of ordinary people, which has resulted in a profusion of red-painted cottages in the Swedish countryside. Actually, this practice gave a triple gain: in addition to its decorative and (originally) status-improving effects, the kind of red paint utilized helped (and still helps) to preserve the wood of the houses.

Artists of earlier times were severely limited in their opportunities for experimenting with colours; the paints and pigments of many interesting and saturated colours were either not available or exceedingly expensive. The situation became drastically altered by the industrial colour revolution in the latter half of the nineteenth century (Section 1.4): the new synthetic stains opened almost unlimited possibilities for colour variations. This has possibly contributed to the more experimental directions taken by art during the last 150 years, including attempts to create art in colours independently of figurative form. Non-representational and colour-rich paintings have often been compared to musical compositions. Such apparent parallels, and their problematic aspects, will be discussed further in the next section.

1.8 Colour and music

Since the onset of colour history, there has been an ambiguous relationship between colour and music. Colours and musical tones both have a kind of direct access to human emotions, albeit with affective effects which may be highly variable and difficult to predict. A great number of more or less original thinkers have been fascinated by the possible parallels between these sensory modalities. The philosopher Locke (1632–1704) published a widely known anecdote about a blind man who knew descriptions of the colour scarlet and concluded that it 'was like the sound of a trumpet'. This has continued to be cited as an interesting analogy; however, the intention of Locke was apparently rather the opposite: he considered the blind man's conclusion absurd and wanted to warn people that one cannot understand colours without actually seeing them.[46]

Already in the ancient world, a musical scale which included the halftones was called 'chromatic' (i.e. 'coloured'), and different musical instruments were said to differ in 'tone colour' (i.e. timbre). Terms like 'tone' and 'harmony' have long been used for both music and pictorial art. Low- and high-pitch tones are often described as being 'dark' and 'light', respectively (or vice versa, see below). Aristotle stated that there were seven main colours, corresponding to the seven tones in an octave, and this tradition continued for hundreds of years until Newton in the year 1704 published his final list of seven colours in the spectrum of sunlight (Section 2.2). Newton arranged the spectral colours in a very useful circular diagram (Plate 2.1c), which is a rather natural way of sequencing the hues because one end of the spectrum is red and the other is violet (Plate 2.4), i.e. a colour perceived as an intermediate between blue and red. Thus, at the end of the spectrum one arrives at something similar to its beginning. As Newton pointed out, this seems analogous to the manner in which tones are perceived along a musical scale, starting off at C and then coming back to a very similar-sounding tone at one octave higher (or lower). Newton's circular colour diagram was directly influenced by a very similar circular graph representing an octave of musical tones, published by the seventeenth century philosopher Descartes. However, in spite of several attempts (see below), nobody has been able to devise a method for calculating precisely which (half)tone corresponds to which colour or hue. In addition, the role of time is very different for music as compared to classical pictorial art: a painting stands where it is, whereas music exists only while being played.

One of the earliest attempts to create a kind of 'colour music' was at the court of Prague by the original Italian painter Arcimboldo (1527–93) who, together with a clavichordist, tried to discover the colour for each tone of their *gravicembalo*. Unexpectedly, these artists considered low-pitch tones to be light and high-pitch tones dark. Unfortunately, it is unclear whether their project ever resulted in a concert performance. Another ambitious and still famous project concerns the *ocular cembalo* of Louis-Bertrand Castel, a French Jesuit befriended by the composer Rameau.[47] Over many years in the 1720–50s, Castel constructed a kind of keyboard instrument in which each key activated some kind of mechanism for the display of a colour above the instrument, e.g. as a coloured piece of paper or as a lamp with coloured glass filters. One version of the instrument was apparently built but it is doubtful whether it was ever used for a public performance.

The development of colour-displaying musical instruments became easier when electric lights appeared in the 1870s. One of the first serious and public colour and tone concerts was staged in 1915 in New York. This was the *Prometheus Symphony* by Skrjabin; the colours of the composition were noted in the score and, during the concert, they emerged on a screen above the orchestra. Later on, performances of various experimental versions of 'colour music' were given in the 1920–30s, partly in the shape of movies with abstract coloured patterns. In the 1920s and later, the Danish-American Thomas Wilfred gave long-lasting public colour-concerts, showing silent patterns of changing colours with no associated sounds. In spite of all these efforts, apparently nobody has yet been able to realize the dream Newton had when he published his seven colours of the spectrum: to use the colours of the rainbow for creating a generally captivating and purely visual kind of music. Musical tones and visual colours remain distinctly different kinds of phenomena.

During the twentieth century, techniques developed that made it progressively easier to produce changing patterns of colour, using various light-emitting devices. Ultimately, combinations of music and simultaneously occurring colour effects became very common (e.g. in discos and the Eurovision Song Contest). However, such cases do not illustrate any inherent parallels between colour and music but simply demonstrate how the effects of two different sensory modalities might be combined to influence our emotions, each modality acting in its own way and according to its own rules and mechanisms.

Particularly toward the end of the nineteenth and beginning of the twentieth century, speculative discussions were common among pictorial artists concerning the possible parallels between colour and music. Many of the nineteenth-century painters saw their palette as something corresponding to the keyboard of a piano. In 1885, Vincent van Gogh (1853–90) attended piano lessons in order to learn more about the nuances of colour tones. The teacher became impatient because Vincent repeatedly interrupted the piano playing for comparisons between the tones he heard and colour samples that he had brought with him.[48] One of the early pioneers of non-figurative painting was Wassily Kandinsky (1866–1944), who thought that colours in some very direct way expressed psychological and physiological states. In his theoretical writings on art, he took much of his terminology from the world of music. For Kandinsky himself, the associations between colour and music were probably very direct and concrete because, apparently, he was a person with marked synesthetic experiences.[49]

1.9 Colour and literature

Within a given culture and country, the use of colours in literary fiction reflects its use in everyday life. Thus, also in literature most or all of the associations between colour and concepts or emotions depend on cultural traditions (see

Section 1.5); they are reflecting nurture rather than nature, and the same colours can be used in widely different contexts. For instance, in an entertaining essay, Staffan Björk (1954) has analysed the role of the colour 'blue' in Swedish literature, and his scrutiny confirms the many-sided and ambiguous nature of colour symbols. Blue heaven for innocence and religious faith; blue eyes for 'golden honesty'; black-and-blue for the results of violent bruising; blue books for parliamentary reports or official transactions; blue mountains and blue birds for infinity and romanticism; American blues for melancholy; blue for conservatism and, in Sweden with its blue-yellow flag, for nationalistic groups and ideas, etc.

Sometimes, colour has been used for expressions describing impossible versions of reality (i.e. *oxymorons*). A well-known French poem by Paul Eluard starts off with the statement that 'The Earth is blue like an orange'. In a Dadaist poem, Kurt Schwitters writes 'Blue is the colour of his yellow hair' (*Anna Blume*). Elsewhere, Gunnar Björling writes that 'I am white like the black cat'.

Several quantitative studies have been made concerning the use of colour terms in various kinds of literary writing. An example of a very detailed investigation of this type is given by the thesis of Alice Pratt (1898). She listed all the colour terms utilized in poems by 17 classical English authors, active in the fourteenth to nineteenth centuries. Her impressive material includes >450 000 lines of poetic text and more than 11 000 colour names. She organized the colour names into six chromatic and three achromatic categories. The chromatic categories red, yellow, green, blue, brown and purple were, in total, mentioned about as often as the achromatic varieties white, black and grey. In all, 52% of the colour terms concerned chromatic categories. Of course, the various authors used a multitude of more or less improvised names for the various colours; in this context, the master of variation and fantasy was Shakespeare, who, for instance, used 50 different words and expressions for the colour of the skin, 30 for different aspects of red and 15 for varieties of yellow. In her conclusions, Pratt emphasized that the association between colour and items of nature was less common for ancient authors than for newer Romantic poets, including English eighteenth- and nineteenth-century classic authors such as Scott, Coleridge, Wordsworth, Byron, Shelley and Keats. A statistical analysis of Pratt's material confirms that those authors born in the year 1700 or later (Romantics) indeed wrote more colourful poems than those of their predecessors. The total number of chromatic and achromatic colour terms per 1000 lines of poetry was more than twice as large for the 10 Romantics (36.1) as compared to seven more ancient writers (13.5).

In addition, her material shows that these two groups also differed with regard to their relative colour choices, i.e. how often each specific colour was named as a percentage of all mentioned colours. As compared to the earlier writers, the Romantics mentioned the colour 'blue' twice as often (9.6% versus 4.7%). For other hues, no such differences emerged between the ancient and newer groups of authors.[50] It seems very appropriate that 'blue' colours have often been used as symbols for romanticism.

Pratt's large amount of statistical material has, almost 100 years later, been re-used by the English researcher McManus (1983) to show remarkable parallels between the literary use of colour names and the differentiation of colour terms within different languages (Berlin and Kay 1969; Plate 1.2; Table 1.1). Results of this kind of meta-analysis are here illustrated in the graph of Figure 1.1, which shows that there is a significant positive correlation between how universal the use of a particular colour category is between languages (cf. 'focus colours';

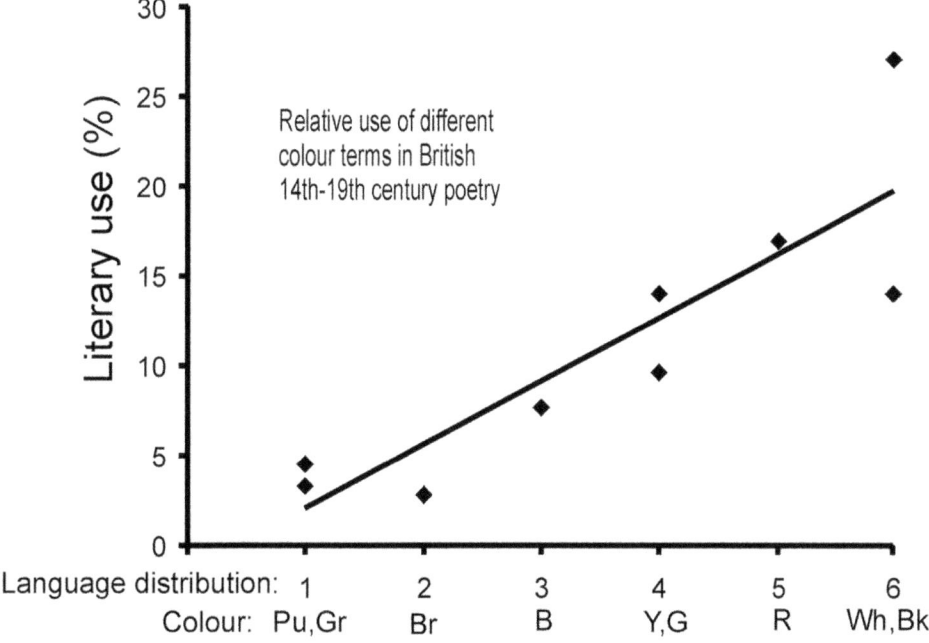

Figure 1.1 Relation between the *Language distribution* of different colour terms vs. their relative *Literary use* in classical British poetry. The x-axis shows ranking numbers for basal colour terms, the rank being higher the more widely used the colour is as focus colour in different languages (cf. Table 1-1; Berlin & Kay, 1969). Abbreviations: *Pu* puple, *Gr* grey, *Br* brown, *B* blue, *Y* yellow, *G* green, *R* red, *Wh* white, *Bk* black. Values along the y-axis show mean values for the relative usage frequency (%) of words corresponding to these basal colour terms in poems published by 13 British poets (data from Pratt, 1898). Terms with a wide language distribution (high x-values) also tended to be more used in the investigated poetry (high y-values). The correlation between these x- and y-values ("*Language distribution*" vs. "*Literary use*") is statistically significant (r = 0.875, n=9, P=0.002); cf. also McManus (1983).

Table 1.1) and how often corresponding colour categories are mentioned in British classical poetry. McManus (1983) showed that this statistical relationship was valid, not only for British poetry, but also for several other kinds of text for which colour names had been counted (e.g. Chinese poetry, modern novels, magazines, popular literature). The mechanisms underlying these associations are still uncertain; more empirical and statistical research is needed within this fascinating field. Perhaps general relationships like those in Figure 1.1 relate to the extent to which various colours are associated with things and events that humankind finds it important to talk about. The highest ranking colours in written texts and language distributions are black and white. Darkness and light, night and day, essential concepts in human life and also directly related to the most important aspects of our sense of vision, the analysis of lightness contrast (see Chapter 4). The highest ranking chromatic colour is red. The red of blood and fire are both associated with some of the most dramatic and alarming colour sensations that people might encounter. Besides, varieties of red in facial colours are of importance for various kinds of human social interactions.

1.10 Colour and philosophy

Philosophers have long been interested in the phenomenon of colour.[51] A relatively recent academic publication edited by Byrne and Hilbert (1997a, b), provides a very wide-ranging review of colour, the first volume concerns exclusively *The Philosophy of Color* (317 pages) and the second volume *The Science of Color* (465 pages). The philosopher's interest in colour is often associated with interest in the analysis and categorization of phenomena related to consciousness. For instance, how can one logically discuss and analyse phenomena that everybody experiences personally, but only purely introspectively and without any possibility of directly observing it in somebody else? Nobody knows precisely how somebody else sees and perceives a colour. We can investigate how colours may be distinguished from each other, but we cannot directly measure the personal colour experience. As a consequence, one of the classic colour problems in philosophy concerns the question of whether a colour in the surrounding world may be thought of as a property of a coloured object or whether it is purely the product of a consciousness.

An entertaining analysis of the relationships between colour philosophy and colour physiology has been published by the philosopher C. L. Hardin in his book *Color for Philosophers* (1993). He effectively neutralizes many of the logical and philosophical classical colour problems by demonstrating how it is possible to discuss colours and colour vision in a very direct and concrete manner, using empirical observations from experimental psychology and neuroscience. That kind of analysis of colour problems will be extensively dealt with in Chapters 2–4. For physiologists, and for philosophers like Hardin, it might even seem self-evident that the existence and properties of colours are wholly dependent on how our retinas and brains react to light entering our eyes. Thus, in this context, it is important to find out how properties of light may be altered by different materials, how different characteristics of light may be interpreted as differences in colour, and how our eyes and brains are equipped for performing this kind of analysis.

2   The signals of colours: light and wavelengths

This chapter will deal with the peripheral mechanisms of colour: which are the characteristics of the physical phenomena that our eyes and brains interpret as 'colours'? As a starting point, it is evident that colours in our surroundings are dependent on light: in darkness we generally see no colours, and the colour of an object can clearly be influenced by the characteristics of the illumination. Thus, for an analysis of colours and colour vision, it is necessary to know more about how varying characteristics of light might influence our sense organs and visual perception. This is a complex subject and the knowledge and insights concerning these matters have been gradually growing during a long period of time. The key concepts concerned are illustrated in a very concrete manner, I think, by considering the progress in this field historically, including a description of some crucial early experiments; this will be done in part of the present chapter.

It is, in this context, important to keep remembering that colours are experiences of conscious brains; as Newton pointed out 'the rays are not coloured'. Colours do not, in themselves, have an independent existence but we see particular colours when lights of appropriate characteristics enter our eyes. Thus, when we talk about a light as being 'blue' or 'yellow', this is a simplified way of saying that the light in question has characteristics that most human observers would perceive and interpret as being 'blue' or 'yellow'.

2.1 Light and darkness: early thoughts concerning the nature of colours

In the ancient world, the sense of vision was interpreted using models which now seem highly exotic. It was believed that, somehow, portions of the observed objects physically entered the eyes and gave rise to the perception. It was sometimes suggested that some kind of rays emerged from the eye itself to cause the correct capture of the necessary 'layers' of surrounding objects. Colours were believed to emerge from mutual interactions between light and darkness. Nowadays, darkness is essentially thought of as an absence of light. In the ancient world, darkness was instead believed to have a positive existence of its own, like some kind of black soup that might be mixed, in various proportions, with the white soup of light. These speculative theories concerning colours and light were not supported by any empirical evidence, but they remained dominant for many hundreds of years, until Newton's experiments with his prisms (see below). Even in post-Newtonian times, ancient colour theories had some famous adherents (e.g. Goethe, Plate 3.4a).

Democritus (~460–370 BC) gave a list of four simple primary colours: white, black, red and *chloron* (pale green). According to his description, other colours could be produced using mixtures of these primary colours. The results of his proposed mixtures are not always understandable for the modern reader; e.g. a blue indigo colour was said to emerge when mixing chloron and black. Later on, Aristotle (384–322 BC) published a somewhat longer list of seven colours, apparently ordered according to their relative lightness: white, yellow, purple-red (crimson), violet, leek green, deep-blue, black. He thought that the colour characteristics depended on the mathematical proportions in mixtures of light and darkness, somewhat analogous to the then well-known mathematics of musical tones and harmonies. Aristotle also suggested that there was a fundamental difference of some kind between 'primary' and 'mixed' colours, and he was very interested in the practical results of colour mixtures, as obtained in the workshops of contemporary painters.

For a long time, until the end of the Middle Ages and the beginning of the Renaissance, the various writings of Aristotle were preeminent in the general understanding of natural phenomena, and this included his colour vision theories. The idea that colours emerged as a result of various combinations of light and darkness were also made widely known by being included in a highly successful encyclopedia of natural science, *De Rerum Naturalis* [*About Natural*

Things] by Bartholomaeus Anglicus; the first edition came in AD 1245 and several different translations were produced until a last printed version emerged in the seventeenth century. This encyclopedia stated that there are 2×7 colours plus the 'primary colours' white and black.

In ancient times, concepts of 'primary' or 'true' colours were sometimes associated with general views on how nature was composed of four natural elements: fire (red), air (blue), water (green) and earth (grey, ash-like). An analysis of this kind was published by the well-known Renaissance artist and writer Leon Battista Alberti (1404–72). He also pointed out that one might produce an infinite number of colours by mixtures of the four 'true' colours plus white and black, i.e. by starting off with only three chromatic colours. He also noted that white and black are not, in themselves, real colours but that their role is to adjust the quality of the true colours, i.e. changing their lightness and saturation. Thus, in contrast to Aristotle, Alberti actually made a distinction between chromatic and achromatic colours.

Nowadays colours are often arranged into diagrams of a more or less circular shape (see Section B.1.1). How old is this method of graphical colour survey and analysis? Circular colour diagrams already occurred in medieval times, and the earliest examples concern the hue variations in samples of urine, used for supporting the diagnostic efforts of medieval physicians.[52] Such graphical displays did, of course, only include colours of interest for this rather specialized application. One of the earliest known examples of a more generalized colour circle was devised by Sigfridus Aronus Forsius (~1560–1624). He was a mathematician, astronomer and astrologer at the Swedish court, and his colour system has achieved a certain degree of posthumous fame after it was discovered in an unpublished manuscript for a book on physics, written in 1611.[53] Forsius did not only arrange colours in circular diagrams, but some of his graphs seem constructed for demonstrating that colours have two independent dimensions: hue and lightness/saturation. Thus, in one of his complex diagrams, four parallel half-circles display the gradual transitions from white to black via each one of four different chromatic colours (red, yellow, green, blue) and, in the middle, similar transitions are shown for achromatic colours (i.e. white to black via grey). In each case the transitions are shown for seven steps, e.g. white–pale blue–sky blue–blue–dark blue–indigo blue–black. The manuscript of Forsius is interesting as an example of early seventeenth-century creativity and lines of thought. However, as it was never published it had no direct influence on colour science. Several decades later, in 1677, a somewhat similar circular system was published by the medical professor

Francis Glisson (~1597–1677) who, like Newton, worked at the University of Cambridge. Glisson's article had the somewhat unexpected title *De Coloribus Pilorum* [*On Hair Colours*].

2.2 White light and spectral colours: Newton's experiments

In 1666, a series of technically simple experiments were performed that would revolutionize colour science. These experiments, concerning the nature of light and colours, were performed by Isaac Newton (Plate 2.1a), then a young man of 23. The observations were made in a dark room, and a narrow ray of sunlight was allowed to enter through a small hole. Newton had acquired a couple of glass prisms at the market of Stourbridge. He let the sun-ray pass through one of his prisms on its way to a white sheet of paper. He then noted that the white sun-ray had been broken up into a characteristic pattern of continuously varying hues, a 'spectrum' (Plate 2.1b). He then proceeded to make two further observations:

1. If one of the coloured light rays of the spectrum passed through a second prism, this did not give rise to rays of new hues.
2. If he re-combined all the variously coloured rays of the spectrum (e.g. using a lens), then white light re-appeared.

The latter two observations are the most important ones in Newton's classical prism experiments. It was already known that prisms might make various colours emerge, and prisms were therefore also sold at general markets (e.g. Stourbridge) as objects for play and entertainment. However, it was generally believed that the prism itself somehow created the new colours. In contrast, Newton's experiment strongly indicated that the white daylight itself contained all the coloured components, that these various components together gave white light, and that the only action of the prism was to separate the various white-light components from each other.

In the context of his prism experiments, Newton also gave the first clear and systematic description of how sunlight is then split up into a characteristic sequence of 'spectral colours' (synonym: 'prismatic colours'). A few years later (1671) he named the phenomenon a 'spectrum'.[54] The bending of the light beam (*refraction*) was the strongest for violet rays and the weakest for red ones. Newton also noted that, within a spectrum, there is a continuous variation of

Plate 2.1 **(a)** Isaac Newton (1643–1727), portrait painted by Godfrey Kneller. © GL Archive/Alamy.

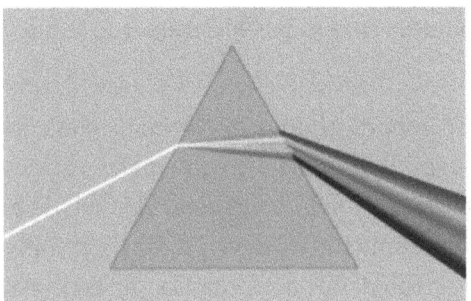

(b) In experiments with glass prisms, Newton showed that the white sunlight contains all the colours of the spectrum (see text).
Source: http://commons.wikimedia.org/wiki/File:Prism_rainbow_schema.png. Image by Joanjoc.

hues. He chose to describe it as a sequence of distinct hues with gradual transitions. However, with this approach it is not self-evident how many distinct hue-steps one should consider for the description. Interestingly, Newton himself used several alternatives in different reports. In an early lecture of 1669, Newton subdivided the spectrum into 11 steps[55] and later on, in his book *Opticks* (1704),

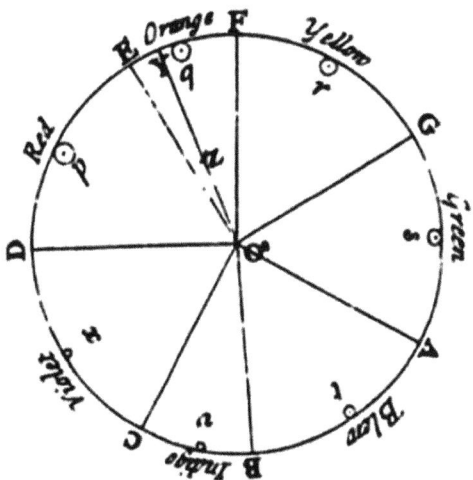

(c) Newton ordered the spectral colours in a circle with white in the middle.
Source: http://en.wikipedia.org/wiki/File:Newton's_colour_circle.png.

comes a description using the seven classical steps: red, orange, yellow, green, blue, indigo, violet. The simplification from 11 to 7 steps was partly done by skipping various red, yellow and green intermediate stages while the 'transitional' colour orange was added. Both lists contain the, for many modern readers, somewhat unexpected blue colour 'indigo'. This extra step of blue was, perhaps, allowed to remain in order to make the final list contain seven colours. As was already mentioned earlier there was a widely spread idea in the ancient world (e.g. Pythagoras, Aristotle) that, like the musical scale, many natural phenomena had seven categories.[56]

Besides mixing all spectral colours to give white, Newton also investigated the effects of mixing only restricted portions of the spectrum with each other. In this context, he observed that some colours may emerge which are not contained in the solar spectrum: the extra-spectral purple colours, obtained by mixing red and violet/blue. Newton summarized some of his findings in a newly devised colour circle (Plate 2.1c), in which the spectral hues occur in the spectral order and in which the circle is closed by connecting the red and violet ends of the spectrum. Saturated colours occurred along the circle periphery, white was in the centre, and various shades of unsaturated colours were in between. Unfortunately, Newton chose not to include the extra-spectral purple colours into his circle; they would have fitted well between red and violet, and they were also placed there by later investigators using similar circles. Newton suggested that his colour circle might be used for predicting the results of colour mixtures: one might draw a line between the two starting colours and use their relative intensities for calculating the position of their combined 'centre of gravity', i.e. the colour resulting from the mixture. Newton's colour circle already includes many of the features of modern chromaticity diagrams, used for the prediction and calculation of additive colour mixtures (cf. Plates 2.6, B.2).

The results of Newton's prism experiments of 1666 were not presented to a scientific audience until 6 years later (1672), and then only briefly. Some of the findings were strikingly anti-intuitive and Newton received an unexpected stream of critical remarks; this was probably one of the reasons why it took so long, until 1704, before he published a full account of these findings in his book *Opticks*. For his contemporaries, the most paradoxical and least acceptable of his findings concerned the composite nature of white sunlight, that it consisted of numerous components of different colours which together became white.

2.3 The rainbow

The rainbow gives a direct demonstration of how white sunlight can be broken up into a spectrum of different hues (Plate 2.2); it is, in a way, a naturally occurring version of Newton's prism experiment. The phenomenon becomes visible for an observer with rainy weather in front and a radiating sun behind his/her back. The sunbeams will be reflected and refracted within the various single droplets, and following a single-stage reflection the rays will emerge from each droplet at different angles according to their spectral colour (Plate 2.3b, right droplet). The angle of refraction is smaller for red than for blue or violet light, and therefore the eye of an observer will see blue/violet rays from droplets that are closer to the ground than those for which red rays are seen (see Plate 2.3c). Hence, the classical 'primary' rainbow has red on the outside (viewing angle 42.3°) and violet on the inside (40.6°). In strong sunlight a weaker 'secondary' bow may be seen outside the primary one (viewing angle 50–53°). In the secondary bow the order of spectral colours is reversed (red inside, violet outside) and the coloured lights are weaker because in this case they are seen after a double reflection within each droplet (Plate 2.3b, left droplet).

As one might expect, Newton gives a thorough and still valid quantitative analysis of rainbows and their mechanisms. He also cites a few earlier seventeenth-century investigators who had understood the importance of intra-droplet reflections for rainbows; these predecessors did not, however, realize

Plate 2.2 Sunshine and rain combining to a rainbow, Nature's own demonstration of sunlight's colour spectrum. A weak secondary bow is seen above the stronger primary version.
Source: http://commons.wikimedia.org/wiki/File:Rainbow_02.jpg.
Image by Jerry MagnuM Porsbjer.

how the refraction of light might lead to the emergence of spectral colours from white light.[57] The diagrams that Newton published in his *Opticks* (1704, 1730) are shown, using a somewhat more modern design, in Plate 2.3b and c.

Rainbows are striking and reproducible natural phenomena of common occurrence, and it is interesting to note how differently they have been depicted, described and interpreted at different times.[58] The differences concern both variable degrees of simplification and alterations in the colours and their sequence. Aristotle described the colours of the rainbow as red–green–purple, which is a simplification but still gives a correct inward-directed colour sequence. Other, more exotic and unrealistic sequences were described by the ancient stoics (sequence: yellow–red–purple–orange–blue–green) and in illustrations from the sixth century (sequence: red–white–green). In medieval illustrations, the rainbow is often shown having only two or four different hues, which is in contrast to descriptions of the ancient Roman poets Ovid and Virgil, who considered rainbows to contain at least 1000 different colours. Inspired by these authors, some medieval pictures do indeed show rainbows with a great number of differently coloured concentric lines. Beautiful and more realistic

(a)

(b)

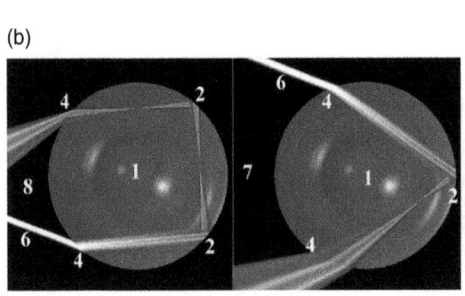

(c)

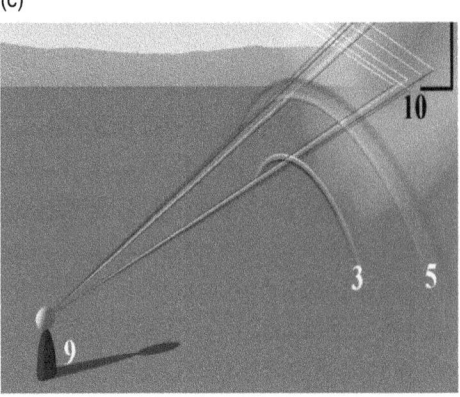

Plate 2.3 (a) Example of rainbow in art: "*The blind girl*" by John Everett Millais (1829–1896), using the rainbow as a symbol for things lost to the blind. (b) Diagram showing how refraction separates sunlight into its different wavelength components when light leaves a rain drop. In a rainbow, the refraction follows an inside reflection which, in each droplet, happens once for the primary (b, right) and twice for the secondary bow (b, left). (c) Diagram showing the observation angles and colour sequences for primary (lower) and secondary (upper) rainbows.

(a) *The Blind Girl* oil on canvas 1854–1856, by Sir John Everett Millais, 1st Baronet, PRA (1829–96). © World History Archive/Alamy.

(b–c) Rainbow mechanisms.

Source: http://commons.wikimedia.org/wiki/File:Rainbow_formation.png.

Image by Peo.

rainbows were painted in the seventeenth century by artists like Rubens and Ruisdael. In the nineteenth century, the phenomenon became a popular motif for English landscape painters like Constable and Turner, who both had an interest in meteorology; Turner has even depicted the less prominent and weakly coloured bow that may become visible in rain illuminated by moonlight.

In the course of the history of art, rainbows have often been used as symbolic instruments, in different ways at different times. For instance, in *The Blind Girl* by the Pre-Raphaelite nineteenth-century artist Millais (Plate 2.3a), the blindness is accentuated by adding two very brightly coloured rainbows that the girl herself could not see.

2.4 Properties of light: wavelength and frequency

One of the early and most prominent post-Newtonian scientists studying the nature of light and the effects of colour mixtures was the Englishman Thomas Young (Figure 2.1a), a multi-talented man who also contributed to the interpretation of Egyptian hieroglyphs.[59] Newton considered light to be a radiation of particles, which is still one of the two valid models; the particles in question are nowadays called *photons*. Young pointed out that light may also be regarded as a travelling wave; he suggested this at about the same time as the French scientist Augustin Jean Fresnel (1788–1827). To support his theory, Young analysed phenomena of multi-beam interference: how the intensities of superimposed beams of light might interact in a manner similar to that of superimposed waves of water. Waves added while simultaneously reaching their peak will result in a larger summed wave, and waves added while 'out of phase' (one low, one high) will diminish each other. A travelling wave may be defined by its speed of progression and wavelength, i.e. the distance between consecutive waves. Newton had already published very detailed measurements of patterns of light interference, concentric light- and dark-coloured bands that occurred close to the contact area between a convex lens and a flat piece of glass. These interference patterns, now called 'Newton's rings', were very cleverly used by Young for calculating the wavelengths of visible light. As one of his results, Young determined the wavelengths for the various portions and colours of the solar spectrum. In a prism, he concluded, the refraction was different for light of different wavelengths and this caused the white sunlight to become subdivided into a spectrum of differently coloured portions (Plates 2.1b, 2.4).

(a)

Figure 2.1 **(a)** Thomas Young (1773–1829), portrait painted by Thomas Lawrence. Using studies of interference phenomena, Young showed that light may be regarded as a wave movement. He proposed that all colours may be seen using only three types of sensors with different wavelength sensitivites.
© Classic Image/Alamy.

(b)

(b) Hermann von Helmholtz (1821–1894) continued the analysis of trichromatic colour vision (the "*Young-Helmholtz theory*").
© nicku/123RF.com.

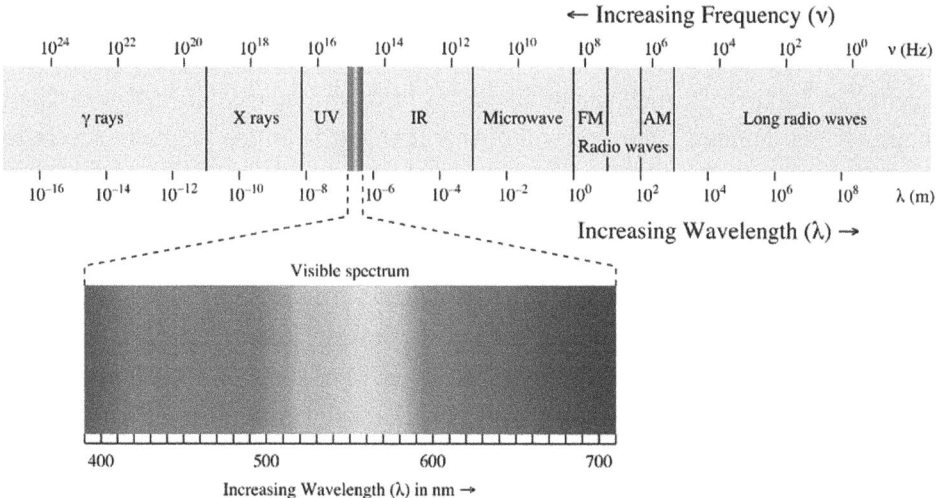

Plate 2.4 Different types of electromagnetic radiation, having different wavelengths (lower scale, m) and frequencies of oscillation (upper scale, Hz). The visible portion of electromagnetic radiation (i.e. "light") covers a narrow region of wavelengths from just under 400 to a little above 700 nm (1 nm = 1/1 000 000 000 m). Adjoining regions are called UV (ultraviolet) for shorter and IR (infrared) for longer wavelengths (i.e. for higher and lower frequencies respectively). The lower portion of the diagram shows the spectrum of sunlight (see sections 2.2 and 2.4).
Source: http://en.wikipedia.org/wiki/File:EM_spectrum.svg.
Image by Philip Ronan.
A black and white version of this figure will appear in some formats. For the colour version, please refer to the plate section.

Another British scientist, whose contributions were of major importance for our understanding of light and colours, was James Clerk Maxwell (see Plate 2.5a). Based on his investigations, he concluded that visible light is merely one example of the more general phenomenon of *electromagnetic radiation*, which may be made to appear by interactions between electrical and magnetic fields. Other low-energy variants are, for instance, the radio waves, which are easily produced using electrical devices. We now know that the phenomenon of electromagnetic radiation covers an enormous range of different wavelengths (Plate 2.4). Besides the measurements of wavelengths, it is also of interest to know how rapidly the waves succeed each other, i.e. the frequency of the wave movement. Both measures are used for defining the type of electromagnetic radiation, and they are easily converted one to the other due to the constant speed of electromagnetic radiation in a vacuum.[60] The energy content of individual photons is directly related to their wavelength and frequency: the shorter the wavelength and the higher the frequency, the more energetic are the photons. In the remainder of my text, I will use wavelength as a measure of light quality.

Biological and other effects of electromagnetic radiation are strongly related to its wavelength. Radio beams (Plate 2.4) have extremely long wavelengths, from metres to kilometres; such radiation passes through bodies and houses without causing any damage, and also without us noticing it unless we have access to a radio receiver. The microwave oven heats our food using wavelengths of centimetres, and also the pleasant warmth of hot non-glowing things is carried by radiation with longer waves than visible light. For instance, for an object of ~27°C (= ~300K) the heat radiation has a peak at wavelengths of 1/100 mm, which is more than ten times longer than the wavelength of red light. The deep red portion of sunlight has its longest visible waves at about 730 nm (1 nm = 10^{-9} m = 1/1 000 000 mm). The shortest waves that we can see are about 380 nm long and look violet. Sunburn is mainly caused by the even shorter-waved ultraviolet (UV) radiation; in our bodies external UV radiation predominantly has an influence on skin and other superficial structures. Still shorter and more energetic waves may, however, pass through the whole body and the photons of such radiation may be dangerously powerful and cause damage when hitting various molecules and cell components. This category includes, for example, X-rays and the gamma radiation associated with radioactive processes.

The borderlines for the section of electromagnetic radiation that we can see are not completely sharp, and there is a certain degree of variation between different normal subjects. In particular for the upper longwave limit, rather different values appear in different accounts, varying from 700 to 780 nm; all these wavelengths are seen as red. In a normal light microscope, structures might be distinguished down to distances of about half the wavelength of the light employed, i.e. for violet light down to about 0.2 μm (1 μm = 1/1000 mm). Light microscopes are good enough for looking at whole cells and large cell components; a red blood corpuscle is, for instance, about 7 μm in diameter. For the observation of small details inside cells, other kinds of radiation than visible light must often be used (e.g. the electron radiation in electron microscopes).

Within the broad range of possible wavelengths for electromagnetic radiation, the visible light covers only a very narrow band (Plate 2.4). This, for us, important band has, however, been 'chosen' in a highly appropriate way during the course of our biological evolution. When arriving at the surface of the earth, the sun's radiation has a very different total intensity for different wavelengths. The superficial temperature of the sun itself is such that its total radiation has its maximal intensity within the narrow band of visible wavelengths (Figure 2.2), i.e. this particular range of wavelengths is very appropriate for the 'illumination' of our earthly surroundings. Furthermore, this part of the

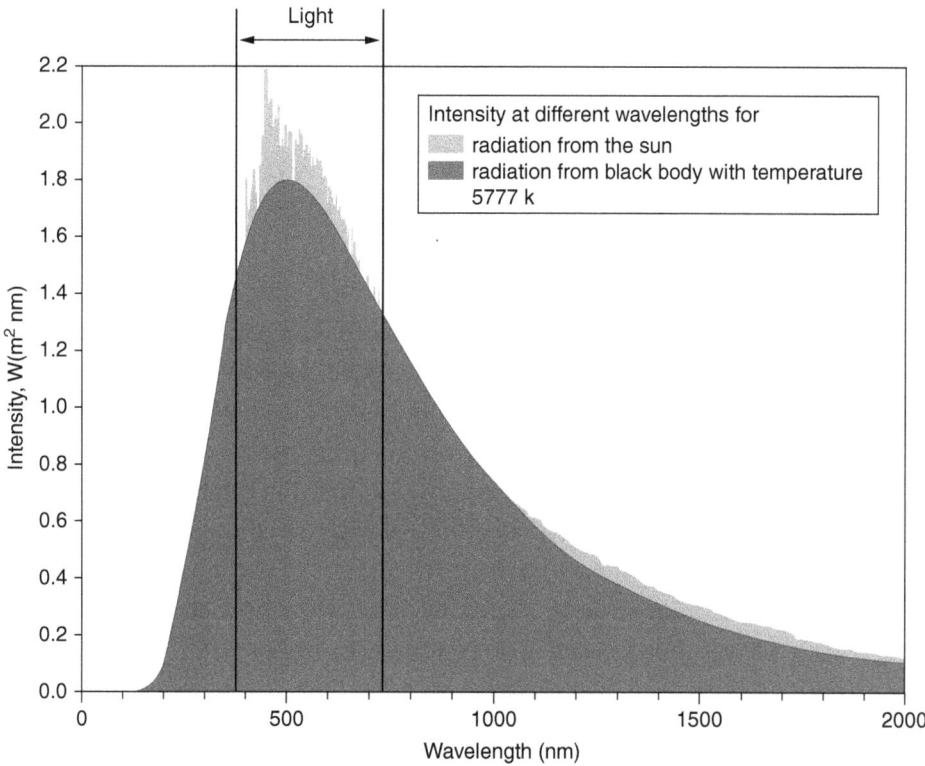

Figure 2.2 Diagram showing the energy content (W/m² nm) for different wavelengths (nm) of sunlight when the beams arrive at the earth's atmosphere. The highest energy content is present within the region of visible light ("*Light*"), from wavelengths just under 400 to those just above 700 nm. The distribution of sunlight-energy across different wavelengths is very similar to that seen for a "black body" heated to a temperature of 5777°K (= 5777 - 273 = 5504°C; see dark-grey area in diagram). Source: http://en.wikipedia.org/wiki/File:EffectiveTemperature_300dpi_e.png Image by Sch.

radiation carries enough energy for the activation of biological processes without being dangerously powerful at the photon level, i.e. it causes no damage to cells and cell components. The risk for cell damage is higher and potentially serious just outside the range of visible light, for the shorter-waved UV radiation.

The very narrowness of the band of visible light is actually advantageous from an optical point of view. As has already been mentioned above, light of different wavelengths will be refracted to different degrees when passing through the same lens or prism, and pictures projected into the eye via its lens system would be much less sharp if the light-sensitive tissue of the eye (the retina) were able to react to a much broader range of electromagnetic radiation than 380–730 nm. Even within this limited range, the wavelength-associated variations in optical refraction constitute a problem requiring various kinds of correction and compensation (see Section 4.3.2).

2.5 Trichromatic nature of colour vision: all hues from three

Newton had shown that mixtures of the various prismatic components of sunlight might give new colours, e.g. mixing them all might cause white, mixing pairs of hues might cause hues to appear other than those of the parents. He summarized such effects with his colour circle, and this line of research was taken up and developed further by several of his successors.

Mixtures of different coloured lights can, of course, be created as in Newton's classical experiments: using a prism for splitting up the sunlight and then superimposing various spectral components using optical devices (e.g. lenses, prisms, mirrors). A more modern alternative (already mentioned above) is to use lamps with colour filters, each one transmitting only a particular range of wavelengths. Mixing is then easily done by projecting light from different lamps onto the same point of a white screen (Plate 1.5b). With a technically still simpler method, additive mixtures of reflected light may be produced using a 'colour top', a rotating circular object with sectors painted in different colours (Plate 2.5b). Provided the speed of rotation is high enough, the colours of the different sectors will merge to produce a single one (cf. Section 1.6). Similar methods were used already in ancient times; e.g. Ptolemaios (second century AD) used a rotating multicoloured disc, mounted on a potter's wheel, for studies of colour mixtures.[61]

Both before and after Newton, many systematic experiments were done concerning the effects of mixtures of coloured pigments or lights, and the conclusion was reached that all hues might be produced using only three starting colours. Inspired by such findings, it was suggested that human colour vision might depend on the functions of three different kinds of light sensors in the eye, each sensor being sensitive to a different quality of light (i.e. a different 'starting colour'). Such suggestions had already been published in 1777 by the glass merchant George Palmer and, a little later (1802), by the scientist Thomas Young (Figure 2.1a). A major difference between these two early hypotheses was that Palmer believed that there were only three distinct kinds of light, while Young understood that visible light contains a continuous variation of wavelengths, and he suggested that each one of the three kinds of colour sensor was sensitive to a particular range of wavelengths. This proved to be largely true

Plate 2.5 **(a)** *James Clerk Maxwell (1831–1879)* performed important basic research concerning the nature of electromagnetic radiation, including light. He analyzed trichromatic colour vision and the effects of additive colour mixtures, including mixtures produced using a rotating "colour top" **(b)**. © nicku/123RF.com.

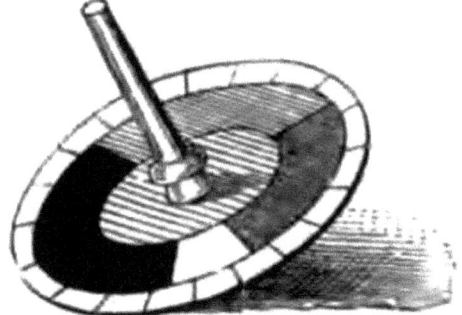

(b) Colour top.
Source: http://commons.wikimedia.org/wiki/File:Color_top_1895.png.

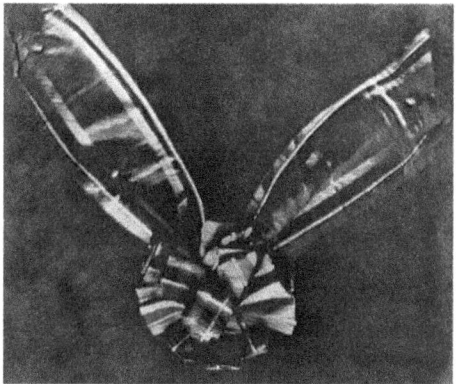

(c) Maxwell also produced the first-ever colour photograph (1861, a coloured tartan ribbon), which then was shown by superimposing three coloured light projections.
Source: http://en.wikipedia.org/wiki/File:Tartan_Ribbon.jpg.
A black and white version of this figure will appear in some formats. For the colour version, please refer to the plate section.

(Chapter 4). At first, Young proposed that the three sensors were maximally sensitive for red, yellow and blue respectively; later on he suggested red, green and violet. These lines of thought were subsequently further developed and experimentally explored by the German scientist Hermann von Helmholtz (Figure 2.1b), and the theory of human trichromatic colour vision was often cited as the 'Young and Helmholtz theory'. However, these two gentlemen never met, Helmholtz was about 8 years old when Young died.

Besides Young and Helmholtz, several other prominent scientists were also engaged in the early analysis of human trichromatic colour vision and in the development of quantitative methods for predicting the effects of colour mixtures. Among others, a series of important contributions were made by James Clerk Maxwell (Plate 2.5a), already mentioned above. He used his observations and insights for producing what has become known as the world's first colour photograph (Plate 2.5c), a picture showing a knotted tartan ribbon (Maxwell was

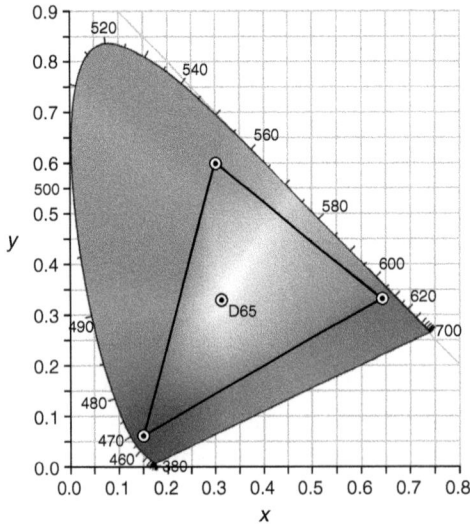

Plate 2.6 CIE-1931 chromaticity diagram, including an indication of the inner triangular region of colours covered by the sRGB-system for colour reproduction. This latter system is widely used and it corresponds to the original gamut for European colour television. Many of the most saturated colours that human eyes can see fall outside the range of the sRGB system. The point labelled D65 shows the CIE-coordinates for the white colour of a standard daylight type of illuminant.
Source: http://commons.wikimedia.org/wiki/File:CIExy1931_sRGB.png.
Image by PAR.
A black and white version of this figure will appear in some formats. For the colour version, please refer to the plate section.
The *xy*-coordinates for the sRGB system: Red 0.64/0.33, Green 0.29/0.60, Blue 0.15/0.06; see Section 11.2 in Hunt and Pointer (2011).

a Scotsman). In 1861, this photograph could only be shown by projecting three component pictures on the same screen, superimposed on each other. Each component picture had been photographed using monochromatic film (nothing else was available in 1861), but with a different colour filter in front of the camera. At the time of projection, corresponding red, green or blue filters were placed in front of each one of the three projectors. This Maxwellian method for colour separation and reproduction is, in principle, very similar to techniques that are still in use for colour printing and photography (cf. Plate 2.7, Section 2.6).

When studying additive colour mixtures, Maxwell (1860) and others found it practical to arrange the various hues in a triangular scheme with the starting colours at each one of the corners. In such a scheme, the hue of a mixture would typically be lying along the straight line connecting two starting colours. Plate 2.6 shows a well-known and more modern version of such a diagram, the 1931 CIE chromaticity chart, so named after its originator, the Commission Internationale de l'Eclairage (the International Commission on Illumination). This diagram also has a triangular shape, albeit with a rounded upper corner; the inner, more pointed triangle represents a technical subsystem of colours (see caption). The various spectral hues are shown in clockwise direction along the outer rim, from 380 nm violet via blue, green, yellow and orange to 700 nm red. The extra-spectral purple colours are shown between the two lower corners (Plate 2.6). As in Newton's colour circle, less saturated colours lie further inwards and achromatic colours are in the centre (for further details on CIE diagrams, see Section B.1.2).

2.6 Trichromatic colour production and reproduction

The discovery that only three starting hues are needed for mixing all the other ones was, of course, technically extremely important when designing procedures for colour reproduction and display. As has already been preliminarily dealt with in Chapter 1 (Section 1.6), there are two main technical types of colour reproduction: (1) making different hues by direct additive mixtures of coloured lights (e.g. in TV and computer screens, projectors, etc.); (2) causing different hues to appear in subtractive mixtures of pigments or paint, meant to be seen with reflected light (e.g. coloured substances applied to paper, textiles, wood, plastic, building materials, etc.). Different starting colours are typically used for these two technical types of colour mixing.

2.6.1 Additive mixtures using coloured lights (RGB)

These mixtures are typically done using a so-called RGB model, as named from the three primary starting lights: Red, Green and Blue. The exact wavelengths of these three primary lights are not very critical; the most important conditions are (1) that none of them may be obtained by mixing the other two, and (2) that they should be situated far from each other within the chromaticity diagram (see inner triangle of Plate 2.6).

In an ordinary TV or computer screen, a great number of small and dot-like light sources are situated very close to each other, each one capable of producing different intensities of either a red, green or blue light when activated. Each such light-emitting dot is called a 'pixel', a term derived from 'picture element'. In a commonly used version of the RGB system (sRGB, Plate 2.6), R corresponds to light with a peak intensity at 610 nm, G at 550 nm and B at 465 nm. For practical technical reasons the pixels of the TV screen cannot produce all of the most fully saturated colours that we might perceive with our eyes. Hence, the colours of the commonly used sRGB system cover only an inner triangular part of the CIE chromaticity diagram (Plate 2.6). The subsets of colour (*gamuts*) for different RGB systems and colour applications (e.g. TV, monitor screens) typically cover variously large triangular regions within the CIE chart;[62] a complete correspondence to the CIE chart (i.e. to all visible colours) is technically difficult to realize.

Plate 2.7 shows an example of how the RGB system is used for the display of a colour picture on a computer monitor, superimposing the three separate pictures created by the R, the G and the B pixels, respectively (cf. Plate 2.5c).

2.6.2 Subtractive mixtures of coloured substances (CMYK)

In colour prints, several transparent layers of pigment are put on top of each other, and for such applications three kinds of coloured ink are used, having absorption properties suitable for subtractive colour mixtures. The most commonly used colour model for subtractive mixing is named CMYK, which is an abbreviation indicating the primary colours used in this context: Cyan (blue-green, absorbs red), Magenta (purple-red, absorbs green), Yellow (absorbs blue), and Key black, an old technical term for a black printer's ink (Plate 2.8). The black ink is used for obtaining deeper blacks and sharper details than would be possible by simply printing the C, M and Y inks on top of each other.

The name 'cyan' comes from 'kyanos', the Greek name for the colour blue. Magenta is an extra-spectral colour, in additive mixtures obtained by red +

Plate 2.7 Example of how a colour picture may be made, using a combination of the R, G and B components of the original.
Source: http://eo.wikipedia.org/wiki/Dosiero:Barn_grand_tetons_rgb_separation.jpg.
A black and white version of this figure will appear in some formats. For the colour version, please refer to the plate section.

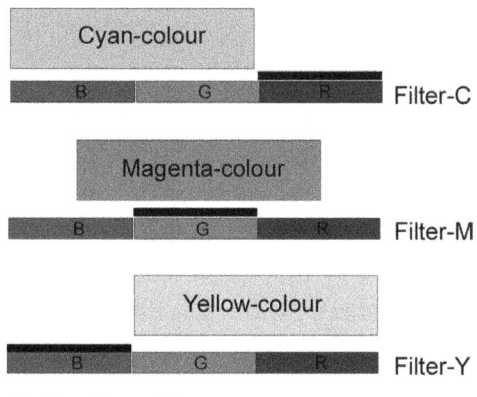

Subtractive mixtures:
 Filter C + M = Blue
 Filter C + Y = Green
 Filter M + Y = Red

Plate 2.8 The starting colours of the *CMYK-system* for subtractive colour mixtures: Cyan, Magenta and Yellow. Each one of these colours will appear if white light is passed through a filter with roughly 1/3 of the spectral wavelengths blocked (see thick black line in filter-diagrams). The K component of the CMYK-system is black. Lower part of the illustration: list of subtractive CMYK-mixtures that give the starting colours for the additive RGB-system. A black and white version of this figure will appear in some formats. For the colour version, please refer to the plate section.

blue/violet (Plate 1.5b). The name was introduced by the English firm that started selling this synthetic colour. This happened in 1859, soon after the Battle of Magenta, and it has been suggested that name of the colour was inspired by all the blood flowing at this battle.

There is a logical correspondence between the RGB and the CMYK models: each one of the three possible combinations of two CMY inks gives, after their joint absorption of about two-thirds of the spectral wavelengths, the remaining range of wavelengths roughly corresponding to one of the RGB colours (Plate 2.8). These are the wavelengths that will be reflected from the printed paper and reach our eyes.

As compared to additive mixtures of light, the results of subtractive mixtures of pigments are more difficult to predict and the outcome might be unexpected and surprising. For subtractive mixtures, the final result becomes darker the larger the number of mixed colours (Plate 1.5a). Hence, for such mixtures, it is advantageous to start off with pigments that transmit a broad range of wavelengths, which is the case for the CMY inks: each one of these inks absorbs only about a third of the spectrum (Plate 2.8). As children, most people learn that with water colours one must mix blue and yellow to get green. However, additive mixtures of these two opponent colours, using coloured lights, produces a very unsaturated colour or even pure grey/white (Plate 1.5b). This classic example illustrates how different the results might become for subtractive and additive colour mixing. Why does the water-colour mixture of blue and yellow become green? This is due to the circumstance that for ordinary water colours (and also for many other kinds of paint) the wavelengths absorbed by the blue plus the yellow pigments usually do not cover the whole middle of the spectrum, and intermediate wavelengths looking green may therefore still become reflected; 'green' covers a very wide part of the spectrum (c. 490–560 nm; Plate B.2b). For a direct additive mixture of blue and yellow lights the result may be white but never strongly green. Furthermore, in additive mixtures the primary yellow and blue lights do not have to cover wide regions of the spectrum in order to produce white; for this purpose, two well-chosen monochromatic single-wavelength lights would suffice (cf. yellow a' and blue b' in Figure B.1).

The first colour prints were made in the 1480s by Erhard Ratdolt, who worked in Venice and Augsburg. He used woodcuts with the images of four separate blocks printed on top of each other for each picture, one block for each mixing-colour.[63] However, most of the coloured illustrations of early printed books were mechanically reproduced in black, and the colours were added afterwards by hand. For the commercial production of true colour prints, one pioneer was the German artist J. C. Le Blon (1667–1741), who started his experiments along this line in 1710 in Amsterdam. In 1717 he moved to London, where he established a Picture Office. Le Blon used metal plates with picture elements engraved according to the mezzotint technique. For each picture, the engravings of 3–4 plates were printed on top of each other, a separate ink being used for each plate (usually blue, yellow and red, sometimes also black). At this time, long before photographic techniques were available, the manual/intuitive separation of a coloured picture into its 3–4 chromatic components, including the preparation of corresponding mezzotint plates, was a very slow and difficult process with many trial-and-error elements. Furthermore, the available inks did not have properties that were optimal for colour printing. Hence, Le Blon had great difficulties in getting sufficiently good results with his printing and he often had to retouch the printed pictures by hand before they could be offered for sale.[64]

2.7 Colour hierarchies: primary, elementary and complementary colours

It is important to realize that the meaning of the term 'primary colour' is context dependent. Early on, this term was mainly used for denoting the starting colours needed for generating all the other ones. A few years prior to Newton's prism experiments, the Irish chemist and physicist Robert Boyle published a book in which he introduced the concept of primary colour for chromatic colours that are efficient as starting colours for subtractive mixtures.[65] His primary colours were red, yellow and blue. In addition, white and black were needed to produce variations in colour lightness and saturation. Both before and after Boyle's book, the preferred starting-colour triad of red, yellow and blue returns in numerous writings and statements concerning subtractive mixtures in painting. As described in preceding sections, present-day printing techniques instead use yellow together with bluish-green (cyan) and bluish-red (magenta) as subtractive primaries (CMYK model, Plates 1.5a and 2.8). For additive mixtures, the modern primaries are typically red, green and blue (RGB model, Plates 1.5b, 2.6 and 2.7).

The term 'primary colour' has also been used in perception psychology for indicating colours with a particular introspective quality ('mental primary colours' or 'elementary colours'). The term signifies hues that normal subjects experience as being unique and which, therefore, can only be denoted by their own name; such hues are also indeed referred to as 'unique' or 'unitary' hues. According to perception psychologists, there are only four chromatic colours with unique hues: red, yellow, green and blue. This class of primary colours also includes the achromatic black and white. The concept of 'unique' hues was introduced by the German physiologist Ewald Hering (Section 3.2.2) and was originally termed *Urfarben*.[66]

Complementary colours appear in pairs, and in additive mixtures they counteract each other such that, with appropriate proportions, the result becomes achromatic, i.e. white or grey. When instead viewed close to each other, without mixing, complementary colours strengthen each other (Section 3.1.1). Classical complementary pairs include, for instance, red versus green and yellow versus blue.

However, complementarities are not the unique properties of a limited number of pairs of elementary colours; each chromatic colour has its complementary colour.[67]

2.8 Light of naturally occurring colours

The combined results of Newton's colour experiments with prisms and Young's analysis of the wavelengths of coloured lights leads to the tentative conclusion that we have a tendency to see chromatic colours when the wavelength composition of the light deviates from sunlight (for exceptions to this rule, see Sections 2.7 and 3.1.3). How do the relevant changes in light composition normally occur? Why is so much of our surroundings coloured?

There are many different physical mechanisms and processes that might be involved.[68] The most important ones, from a practical point of view, include changes in the composition of the illuminating light (e.g. sunlight) due to:

- reflection of the illuminating light after selective absorption of particular wavelengths. The absorption characteristics depend on the chemical composition of the reflecting material; this is a very common colour-generating process;
- refraction of light in transparent materials;
- interference and scattering of light by materials with an appropriate (surface) structure.

Alternatively, light perceived as being coloured might be actively generated by various processes such as:

- heating;
- electrical activation of gaseous molecules;
- radiation which itself causes a secondary radiation to appear from certain kinds of materials (fluorescence, phosphorescence);
- chemical processes (chemo-and bioluminescence).

These various possibilities are treated in somewhat greater detail below.

2.8.1 Absorption colours: selective absorption and reflection

As sunlight hits an object, three things might happen, to various degrees, for its different wavelength components:

1. They might pass through the material, coming out on the other side (*trans-mission*); if this happens for all wavelengths the object might become almost invisible (e.g. non-reflecting glass).
2. They might disappear into the material in question (*absorption*; Plate 2.9), due to interactions between the light and atoms in the material. A simple form of general interaction is heating, but light may also cause many different kinds of more specific processes to occur (e.g. photosynthesis, see Plate 2.12d). If all wavelengths are absorbed equally the object will look grey or black; if some wavelengths are more absorbed than others the object might look coloured (Plate 2.9).
3. The remaining light, which is neither transmitted nor absorbed, will become *reflected*, which makes it possible for us to 'see' the object (Plate 2.9). Our capacity for vision and colour perception is based on a very complex analysis of how strongly and how selectively the illuminating light is reflected from various kinds of matter. Materials which reflect all the wave-lengths equally and also scatter the light in all directions will look white or grey. If this general reflection takes place without scattering we have a mirror. Materials differ considerably from each other with regard to their reflectance, i.e. how effectively they reflect the light. For white paper the reflectance is about 85% and for black paper about 4%.

Often more than one of these three kinds of light behaviour take place simul-taneously and to different degrees for different wavelengths. The most common reason for the different colours of objects around us is selective absorption: some wavelengths are absorbed more than the others, and therefore the

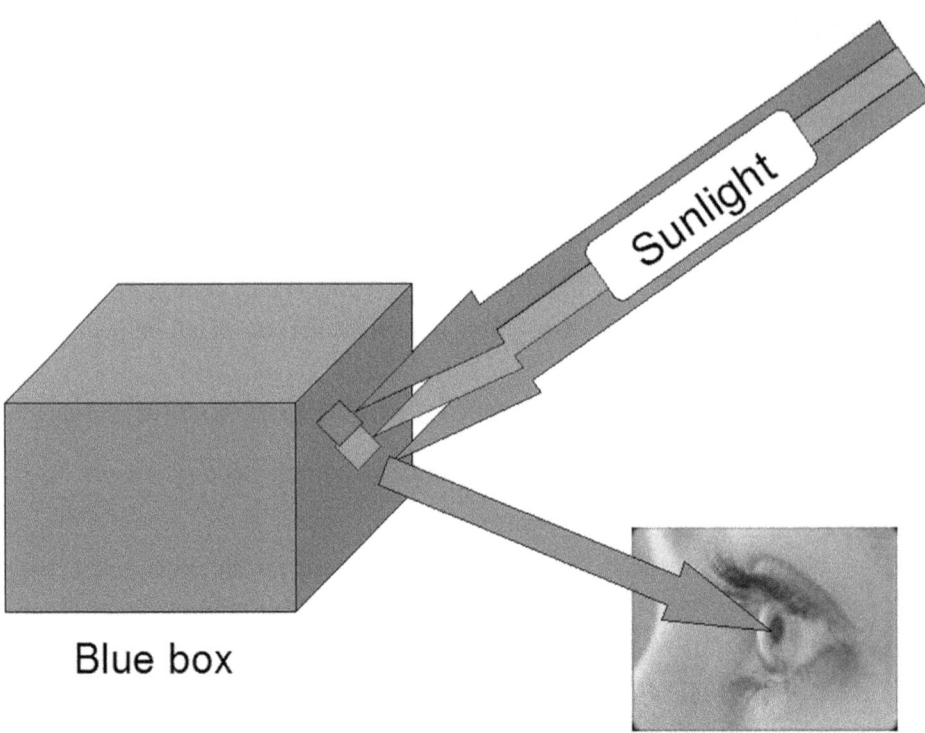

Blue box

Plate 2.9 Diagram illustrating a typical example of why a sunlit blue box looks blue. As the rays of sunlight arrive at this box, those of longer wavelengths disappear into its blue paint (absorption) and only the shorter wavelengths become reflected and reach the eye of the observer; these shorter wavelengths are perceived as being "blue". A black and white version of this figure will appear in some formats. For the colour version, please refer to the plate section.

remaining reflected light will have a different wavelength composition than the illuminating light (e.g. sunlight). If we look at a blue box illuminated by sunlight (Plate 2.9), it looks blue because the longwave components of the sunlight were absorbed into pigments of the box surface, and only the remaining shortwave components were reflected to reach the eyes. The selective absorption of certain wavelengths depends on the chemical composition of the illuminated surface, and colours appearing due to such mechanisms (Plate 2.9) have therefore sometimes been referred to as 'chemical colours' (e.g. terminology of Goethe). In this book, I will call them 'absorption colours'. Why does such a wavelength-selective light absorption take place?

At the level of individual atoms, the absorption or emission of light occurs in a highly wavelength-specific manner. Electrons within an atom can only exist at a limited number of different energy levels. If they are made to move from a higher to a lower level, the amount of energy released corresponds precisely to this particular jump. The energy is released as a light quantum (photon) with a wavelength and oscillation frequency precisely corresponding to the electron jump. This process is reversible: if light quanta with precisely the required energy content are added (i.e. light with precisely the right wavelength), then particular electrons may be made to jump from their lower to their higher energy level. These processes can be directly demonstrated using a gas containing only one species of atoms. Electrical activation of the gas leads to the emission of light at one or several very narrowly defined wavelengths (i.e. at very specific energy levels), and the pattern of such 'emission lines' (Plate 2.10) is highly specific for each kind of atom. Furthermore, for a particular species of atom, light passing through the gas will be absorbed at the same narrowly defined wavelengths as those of the emission lines. Measurements of the pattern of 'absorption lines' can be used to identify which kinds of atom are present in a gas, and this can even be applied to determining the composition of the atmosphere of distant planets and stars. For instance, in precise recordings of the spectrum of radiation from our own sun, a number of black absorption lines ('Fraunhofer lines') appear which reflect the effects of atmospheric light absorption, including that due to the atmosphere of the sun itself. In chemical laboratories, the composition of fluids is often analysed and measured using instruments called *spectrophotometers*: light is sent through the fluid and the degree of its absorption (i.e. the relative light weakening) is measured for different wavelengths.

When measuring the absorption of light into composite and naturally occuring materials the situation is, of course, very much more complex than in studies of homogeneous gas samples: naturally occurring materials usually contain mixtures of a great number of different kinds of atoms and molecules. Therefore, in such materials the absorption of light is typically not limited to only a few distinct wavelengths (no clear absorption lines), but it is instead more gradually varying over broad spectral ranges (cf. Plate 2.10 versus Plates 2.12d and 4.3b). Correspondingly, the intensity of light reflected from natural materials (i.e. the light remaining after the selective absorption) will also typically vary in a gradual manner across broad ranges of different wavelengths (e.-g. Plates 2.12c, 2.13b, c). Within a prismatic spectrum, the separate hues are - *monochromatic* (Plate 2.1b), i.e. each one is caused by effects on the eye of only one wavelength of light (in practice: a narrow range of wavelengths). Light of 580 nm wavelength looks yellow, 600 nm is orange, 720 nm red and 480 nm blue. However, colours emerging due to the selective absorption of light in

(a)

hydrogen

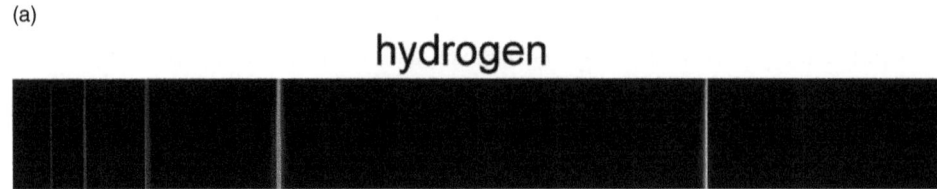

Plate 2.10 (a) The coloured vertical lines show how monochromatic light of different wavelengths is emitted from different kinds of gas (*emission lines*), when activating the gas using electricity. In this illustration, only the most intense and clearly visible emission lines are shown; totally there are many more lines per chemical element. Scale of wavelengths made by the author, using the emission lines. (a) Molecular emission lines for hydrogen (H).
Source: http://commons.wikimedia.org/wiki/File:Visible_spectrum_of_hydrogen.jpg.
Image by Jan Homann.

(b)

helium

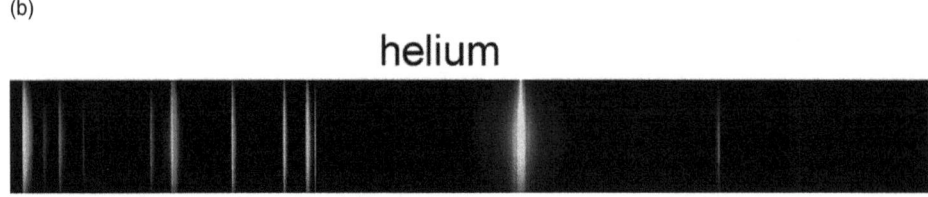

(b) Molecular emission lines for helium (He).
Source: http://commons.wikimedia.org/wiki/File:Visible_spectrum_of_helium.jpg.
Image by Jan Homann.

(c)

mercury

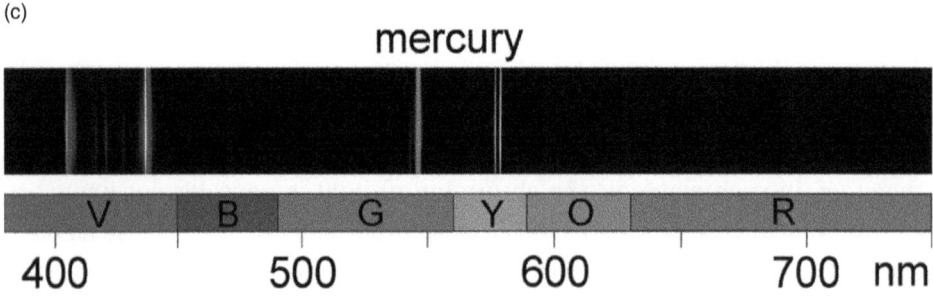

(c) Molecular emission lines for mercury (Hg).
Source: http://commons.wikimedia.org/wiki/File:Visible_spectrum_of_mercury.jpg.
Image by Jan Homann.
A black and white version of this figure will appear in some formats. For the colour version, please refer to the plate section.

Many different combinations of wavelengths can produce exactly the same colour (metamerism).

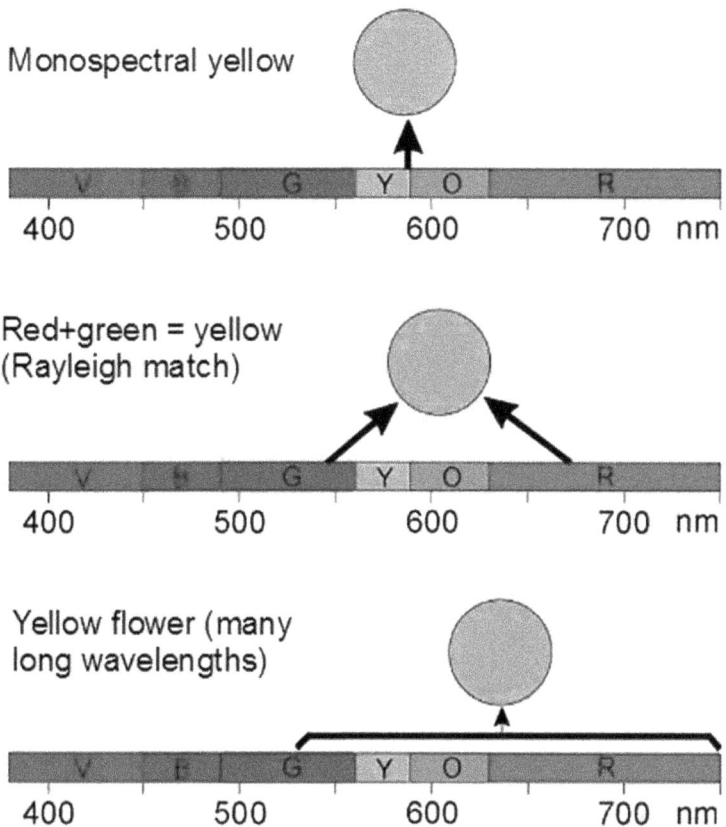

Plate 2.11 Metameric colours: different ways to yellow. A black and white version of this figure will appear in some formats. For the colour version, please refer to the plate section.

natural materials are almost never monochromatic; the coloured light reflected from a piece of such matter consists of a mixture of lots of different wavelengths which, taken together, are interpreted as a single colour by our eyes and brains. We may perceive exactly the same single colour due to a monochromatic light, a mixture of two wavelengths, or a mixture of an enormous number of different wavelengths (Plate 2.11). For our eyes and brains, colours depend only on the relative degrees of activation of three different kinds of wavelength-sensitive receptors in the retina (i.e. the cones), irrespective of the number of wavelengths involved. From some points of view, it is of practical value that most of the

(a)

(b)

Plate 2.12 The green colour of plants. Leaves from elm **(a)**, aspen **(b)** and birch (no photo) are analyzed in the diagram of **(c)** with regard to the reflected wavelengths of light. The measurements of **(c)** were performed using a reflectance spectrophotometer (section 2.8.1). **(d)** Example of how light of different wavelengths is absorbed by chlorofyll a and b. Note that the peaks of light absorption in chlorofyll (i.e., light used by the plant) coincide with regions of low light reflectance from the green leaves.
Source: http://en.wikipedia.org/wiki/File:Chlorophyll_ab_spectra-en.svg.
Image by Aushulz.
A black and white version of this figure will appear in some formats. For the colour version, please refer to the plate section.
Chlorophyll *a* is the most common type of photosynthetic pigment, present in every plant that performs photosynthesis.

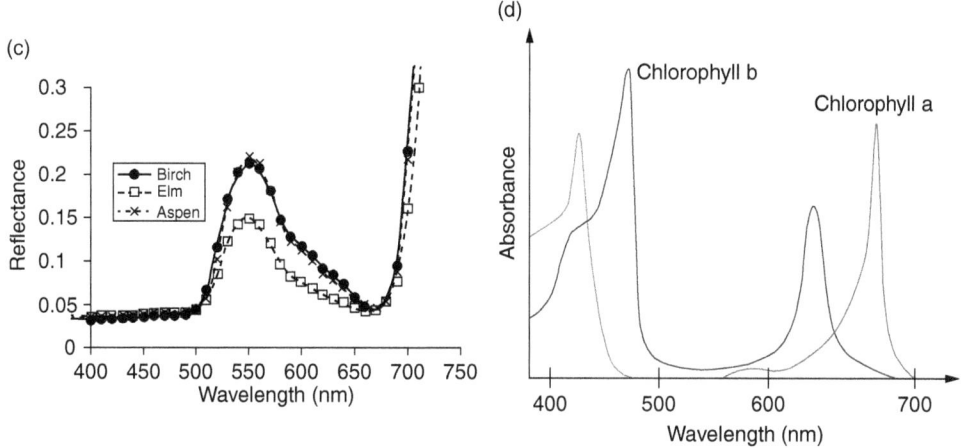

Plate 2.12 (*cont.*)

colours in our natural surroundings depend on mixtures of many wavelengths; if they were all monochromatic the general intensity of the reflected light would be very low.

Much progress has been made with regard to methods for calculating the colour perception resulting from a given mixture of wavelengths and intensities (Section B.1.2). However, the opposite kind of prediction is much more difficult: it is not possible to conclude precisely which wavelengths of light were involved for a given colour perception. One may only make a number of rather coarse guesses, e.g. that, in addition to whatever else is present in the mixture, a red colour would be likely to have much of its light intensity at long wavelengths, and for a blue colour short wavelengths would instead be expected (however, exceptions do occur, see Section 3.1.3). The actual distribution of light intensity across all the different wavelengths cannot be seen; its evaluation requires a measuring instrument such as a spectrophotometer.

For some examples of natural absorption colours, the variation of light intensity across all the visible wavelengths is illustrated in Plates 2.12c and 2.13b–c. In these cases, the intensities of light reflected from leaves and flowers have been measured for different wavelengths, using a *reflectance spectro-photometer*. The instrument employed used a brief flash for illuminating a small area under its lens, and the device then measured the intensity of the reflected light at 36 different wavelengths, covering a range from 380 to 730 nm. A white paper reflected all wavelengths at much the same intensity. Leaves with colours clearly perceived as green (Plate 2.12a–b) indeed reflected much light for wavelengths at the green and yellow middle region of the spectrum but also (counter-intuitively) from several of the longest wavelengths at the red end of the spectrum (Plate 2.12c); this was characteristic for all the many green leaves that I have measured in Småland. The reflected wavelengths are those that were not absorbed, i.e. for green leaves it would include wavelengths that were unimportant for photosynthesis. Hence, one might expect that chlorophyll, which contains the photosynthetic machinery of green leaves, would predominantly absorb light at wavelengths below and above the green middle region of the spectrum. Separate measurements have shown that this is indeed the case (Plate 2.12d).

Plate 2.13b shows a characteristic example of the light composition for natural yellow: petals of the dandelion reflected the light most strongly for a broad range of the longest spectral wavelengths. This is in contrast to the typical results for a blue flower, the blue *Campanula* bellflower (Plate 2.13c); however, in this latter case the reflected light did not only have the expected high intensities for shortwaved portions of the spectrum but also for a considerable range of longer red wavelengths. This was characteristic for almost all the blue flowers that I have measured, including flowers that my wife perceived as being clearly blue.

2.8.2 Structural colours: refraction, scattering, interference

2.8.2.1 Light refraction

In earlier sections, examples were given and discussed as to how refraction might cause white light to become separated into coloured components (Newton's prism, Plate 2.1b; rainbows, Plate 2.3b). This mechanism contributes to the generation of colours in many natural situations. Sometimes the result is seen as beautiful and decorative (e.g. rainbows), occasionally the effect is found to be disturbing (e.g. chromatic aberration in photographic lenses).

2.8.2.2 Wavelength-related light scattering

As sunlight passes through the atmosphere, it will come across numerous air molecules of, in particular, nitrogen and oxygen. Collisions between light and various objects may cause individual light quanta (photons) to change their direction (scattering), and for collisions with objects smaller than the wavelengths of light, like air molecules, this scattering effect is greater for shortwave photons than for light of longer wavelengths. This kind of wavelength-dependent

(a)

(b)

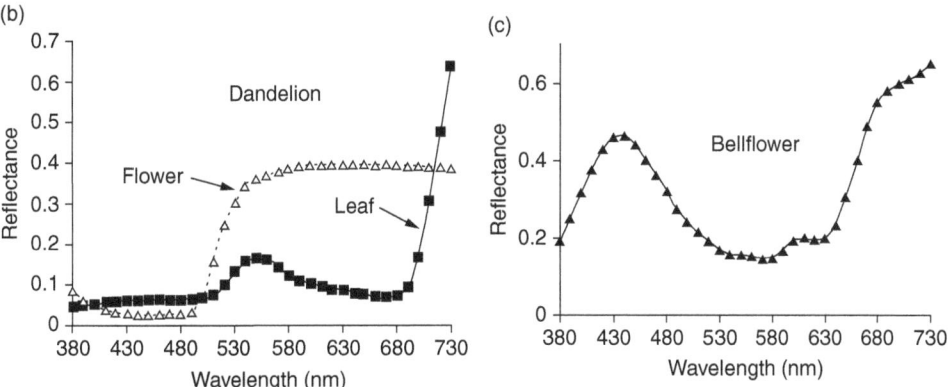

Plate 2.13 (a) Dandelion flowers and leaves on a lawn. (b) Spectrographic measurements of the light reflected from such flowers and leaves (cf. Plate 2.12 c). (c) Corresponding measurements for blue bellflower (*Campanula*; cf. Plate 4.2). A black and white version of this figure will appear in some formats. For the colour version, please refer to the plate section.

scattering is called 'Rayleigh scattering' (discovered by Lord Rayleigh, Figure 5.1b). Due to these mechanisms, sunlight passing through the atmosphere will be more effectively scattered for its (green-)blue-violet wavelengths than for its longer reddish wavelengths, and this causes the sky to look blue in daylight. For similar reasons, distant mountains look blue because they are seen through a thick layer of air with shortwave scattering of the transmitted light. In quite a different context, Rayleigh scattering of light also causes the iris of the eye to look blue when it lacks pigmentation; in this case the shortwave light rays are being scattered by small and colourless protein molecules.

The setting sun looks redder than its midday appearance; this effect is also due to Rayleigh scattering. As compared to the daytime situation, the setting sun's rays will pass through a thicker layer of atmosphere before they reach the human observer. During its longer passage through the air, Rayleigh scattering will cause the directed rays of the setting sun to lose more of their shortwave components, and direct rays ultimately reaching the observer will therefore look redder than in daytime, more dominated by their longwave components.

2.8.2.3 Random light scattering and reflection of all wavelengths together

Crushed glass looks white and is no longer transparent; this is due to the fact that all the little glass grains will refract and reflect the incident light in all directions. In this case we are dealing with objects which are much larger than the wavelength of light, and the scattering then concerns all the wavelengths to about the same degree. A white 'structural' colour of this kind is very common in nature, e.g. for snow, clouds, fog, white feathers, milk, etc.

Sometimes a structural white colour is added to colours due to other mechanisms, e.g. selective light absorption. Consider, for instance, the differences between glossy and rough printing paper. The printing dyes will cause part of the illuminating light to be selectively absorbed, and the remainder will be reflected and interpreted as colours by our eyes and brains. However, some of the illuminating light will be reflected already in the most superficial structures of the paper. For a rough paper (e.g. normal copying paper), irregularities of the superficial structure will cause much of the illuminating light to be scattered in all directions, thereby adding white light to the absorption colours of the printing dyes, making these latter colours look less saturated. A glossy paper has a smoother surface and causes less general light scattering. The absorption colours of the printing dyes will therefore, on a glossy surface, look much stronger and more saturated. For glossies, the light reflected at the most super-ficial layers of the paper goes mainly in one direction, as in a mirror. Hence, by holding the glossy paper at an appropriate angle under the light source, the general reflection of the illuminating light can be largely separated from the light of the printed picture.

2.8.2.4 Interference

Interference between the waves of adjacent light beams may cause the summed intensity to wax for some wavelengths and wane for others, causing a coloured pattern for the observer (see Section 2.4). Such interference phe-nomena may appear when incident light is reflected against surface structures with repeated small irregularities at distances similar to the wavelengths of light. This may cause adjacent light beams to become reflected at slightly different distances and, following their reflection, to become added while oscillating out of phase with each other, which leads to variations of their summed intensities. The phenomenon is sensitive to the viewing angle, which gives interference colours a 'living' and shimmering metallic character. One may easily observe this phenomenon by holding the recording side of a CD under a lamp, turning it back and forth at various angles (Plate 2.14d). Structures giving rise to interference colours are fairly common in animals, e.g. insect wings (Plate 2.14c) and avian feathers, then often in combination with various absorption colours. Other objects displaying interference colours include, for instance, soap bubbles and thin layers of oil on water. Interference mechanisms are also of practical use for subdividing light into spectral com-ponents, using a so-called *diffraction grating*. This is, in principle, a reflec-ting plate with a pattern of furrows at specific and minute distances from each other.

2.8.3 Radiating colours: the active production of coloured light

2.8.3.1 Heat radiation and fire

When heating an object, molecular movements will increase and the object will emit a 'black-body radiation', an electromagnetic radiation with a distribution of wavelengths dependent on the temperature (below also referred to as 'heat radiation'). The wavelengths of this radiation get shorter and,

(a)   Fluorescence   (b)   Bioluminiscence

(c)   Interference (partly)   (d)   Interference

Plate 2.14 (a) *Fluorescence*, light emitted by a material when illuminated using another "activating" light. The illustration shows the coloured fluorescence of various minerals when illuminated with ultraviolet light. The fluorescence phenomenon was named after the mineral fluorite (labelled "F"). (Photo: Hannes Grobe (Wikipedia)).
(b) *Bioluminiscence*, light emitted due to chemical processes in a living organism, in this case a European glowworm. (Photo: Wofl (Wikipedia). **(c-d)** *Interference*, coloured light may emerge due to interactions between different wavelength components of white light, e.g. after reflection in surfaces with an appropriately repeated microstructure. Interference colours change with the angle of viewing and, therefore, they tend to get a shimmering and glittering character (cf. colours of the recording side of a CD). (Sources: see p.xiv. Photo (c): Gregory Phillips, (d) Ubern00b (Wikipedia).)

correspondingly, the energy content of its photons becomes greater the higher the temperature. Our bodies emit a heat radiation at long infrared wavelengths of about 10 000 nm, which are invisible to the eye but may be observed using thermographic cameras. At temperatures exceeding several hundred degrees Celsius, a weak red glowing light becomes apparent, and at 5777K the light has the same wavelength composition as midday sunlight. At still higher temperatures the light gets even more shortwaved, becoming more bluish.

In a burning fire, the flame occurs due to chemical reactions giving a marked increase of temperature, typically caused by a lively oxidation of organic material (i.e. carbon compounds). The colours of the flame depend both on its temperature (i.e. black-body radiation) and on several other factors, including spectral line emissions from various chemical substances (cf. Plate 2.10). In general, the colours within a flame vary with relative temperature: the least hot parts tend to be red and the hottest portions approach white.

2.8.3.2 Fluorescence and phosphorescence

Fluorescence occurs when a material itself emits light as a result of being illuminated. The primary activating light has shorter wavelengths (i.e. higher photon energy contents) than that of the secondary fluorescent radiation. The fluorescence may often be activated using invisible ultraviolet radiation; in such cases, our eyes will see only the secondary fluorescent radiation emitted by the material itself. Earth compounds often have fluorescent properties, emitting light of different colours for different minerals (Plate 2.14a). The phenomenon got its name from the mineral *fluorite* (fluorspar, calcium fluoride; see stones labelled 'f' in Plate 2.14a). Fluorescence phenomena are widely used in association with various analytic techniques in biology and chemistry. For instance, fluorescent substances may be used for labelling interesting cells and structures, making them visible when illuminating the tissue with ultraviolet light.

At night, we compensate for the absence of the sun by using various artificial light sources. Old-fashioned kinds of illumination, e.g. candles or incandescent light bulbs, give a light that is not as 'white' and colourless as midday sunlight (see Appendix C). Fluorescence has become widely used in modern lamps for producing a more sun-like lighting (and also for producing cheaper light). In fluorescent lamps the inside of the glass is covered with a fluorescent substance called 'phosphor' (a name confusingly similar to that for the chemical element phosphorus), and electricity is used to cause an internal mercury vapour to emit UV light, which in its turn activates the fluorescence. The light emitted by fluorescent tubes and lamps often looks rather daylight-white to the eye but, largely due to the internal activating mercury light, it commonly contains a couple of narrow peaks of high intensity at some of the wavelengths for visible light.[69] Colour problems related to illumination will be further dealt with later (Sections 3.1.3 and Appendix C).

For certain kinds of substances, the fluorescence may be directly activated by electricity (*electroluminescence*). Depending on the type of fluorescent substance, different colours may be produced, and there are several varieties of these techniques. Using activating electron beams, electroluminescence is important in devices using cathode-ray tubes, e.g. in old-fashioned oscillo-scopes, computer monitors and TV sets. Another kind of electroluminescence is represented by the light-emitting diodes (LEDs), which are now being developed and introduced into various sections of the consumer market. The LEDs produce light without much heat and they are promising candidates for many general lighting tasks in the future.

Part of the definition of fluorescence requires that the phenomenon should last only 100 ns or less after the end of the activating radiation (1 ns = 1/1 000 000 000 s). A secondary radiation lasting longer is called *phosphores-cence*, and the post-activation duration of phosphorescent light may vary between a few milliseconds up to seconds, minutes or even, in some cases, several hours. Extreme durations of phosphorescent light, e.g. as used in wrist watches, requires a rather long charging period with the primary activating radiation lasting 30 min or more. One of the substances used for such long-lasting non-radioactive phosphorescence is strontium aluminium oxide.

The activation of fluorescence or phosphorescence may also be achieved using radioactive radiation. Before the health hazards of radioactive processes became known, highly active substances like radium were often used as an activator of fluorescence, e.g. in watches. In as far as this general kind of technique is still used, it is only applied with very weak sources of primary radiation (e.g. the hydrogen isotope tritium[70]).

2.8.3.3 Photochemical processes, bioluminescence

Various kinds of chemical processes exist which produce light that is not dependent on simple black-body heat radiation. For instance, white phosphorus emits a blue-green light when oxidized in air, and the less poisonous red variety of phosphorus emits light when mechanically heated by rubbing. The words *phosphorescence*, *phosphor* and *phosphorus* are all derived from the Greek words for light (*phos*) and carrier (*phoros*).

Light produced by chemical processes in living organisms is called *biolumin-escence*. The phenomenon is common in, for instance, fireflies and glow worms. The animal shown in Plate 2.14b is the common European glow worm (*Lampyris noctiluca*), which is actually a kind of beetle, related to the fireflies, which mainly lives on a diet of snails. The emitted yellow-green light helps the males to find the females. Both sexes may emit light, but only the males can fly. The molecule producing the light is called 'luciferin'. The process is highly effective: <20% of the total energy cost for light production disappears as heat.

Bioluminescence is particularly common among marine organisms. At extreme ocean depths, not reached by sunlight, about 90% of all the kinds of living organisms display some form of bioluminescence. This indicates that visual communication and eyes may be important also within a very dark world. However, marine bioluminescence also occurs in more superficial waters. In summertime evenings along northern coasts (e.g. west coast of Sweden), moving the waters might produce a bluish-white light, emitted by minute planktonic animals when disturbed or threatened.[71] Also in these cases, the light is emitted by a kind of luciferin, similar to that of the glow worms.

3 Colours and viewing conditions: not only local wavelengths

3.1 Colour interactions within a visual image

In the preceding chapter, results were described indicating that one might predict the perceived colour of a light if its wavelength composition is known. However, this is mainly valid for small and isolated portions of our central field of vision, e.g. for the observation of a small rotating colour top against a neutral background, or for Newton's prismatic spectrum displayed in a dark room. Our visual surroundings are typically much more complex; normally our eyes will simultaneously encounter a multitude of objects, differing in colour and lightness. Within such a complex scene, the colours perceived for each one of the various details will partly depend on local wavelength compositions, but partly it also depends on complex interactions between the various components of the scene. While we are looking, the brain uses a complex set of rules for correcting and adjusting all the colours before we perceive them.

3.1.1 Simultaneous contrast

In experiments concerning colour perception, colour samples are often used in the shape of small discs or pieces of paper which are shown against a neutral background. By such means, each colour may be judged on its own, in a state of relative isolation from other colours. It is important to realize that the same colour sample, having the same wavelength composition of its reflected light,

Plate 3.1 A drastic example of effects of simultaneous contrast: the two squares with coloured circles have exactly the same darkness and colouring, and this is also true for the upper vs. the lower coloured circle. This "checkershadow illusion" was originally devised by Edward H. Adelson.
Source: http://en.wikipedia.org/wiki/File:Optical_grey_squares_orange_brown.svg.
A black and white version of this figure will appear in some formats. For the colour version, please refer to the plate section.

might give quite a different colour perception in other surroundings or with another background (Plate 3.1). This concerns all the main dimensions of the colour: its hue, saturation and lightness. One of the common kinds of interaction concerns the 'strengthening' effects of adjacent opposites. A grey paper looks whiter against a black background. A yellow paper looks even more yellow against a blue background, and a blue colour becomes stronger against yellow. An interesting and paradoxical aspect of these interactions concerns the behaviour of complementary colours: they strengthen each other for adjacent surfaces, but they may instead sum and become very unsaturated if the two samples are small and close together (e.g. for minor stones in a mosaic, for threads in a woven tissue; cf. definition of complementary colours, Section 2.7).

One of the early scientists studying these colour interactions was the unusually long-lived French biochemist M.E. Chevreul (1786–1889). He is considered one of the founders of modern organic chemistry, and his work in this field led to a renewal of the soap and candle industry. He was appointed head of the dyeing department at the famous tapestry factory of Les Gobelins in Paris, and in this capacity he also isolated and analysed a variety of dyes derived from plants. In tapestries, simultaneous colour contrast at the level of the details (e.g. the various interwoven threads) is of great importance for the visual impression of the whole depicted scene (cf. Plate 1.6c). These kinds of problems were systematically analysed by Chevreul, and in 1839 he summarized these investigations in a very influential book, which included studies of colour contrast, colour harmony, and other related subjects.[72] Observations were also described concerning additive colour mixtures with small spots of different colours placed close to each other (cf. pointillism, Plate 1.6b); Chevreul's book was important for the thoughts and work of contemporary artists like Delacroix and the French impressionists.

The size of a painted picture is also of importance for colour perception. In a very small picture, the colours may look weaker (a problem for colour-blind persons, see Section 5.5.2). Therefore, the miniature illuminations of medieval handmade books are often in colours which might have looked too strong and vulgar if they had been applied in a monumental wall painting.

3.1.2 Negative and positive afterimages

Besides the simultaneous contrast phenomena there are also 'serial' contrast effects, i.e. colour contrast with a time factor. These effects are seen in *negative afterimages*, visual perceptions that appear *after* looking at a scenery or picture.

The appearance of a *negative afterimage* requires that one has first steadfastly stared for 10–20 s at the target picture and then directed the eyes toward a neutral non-structured item (e.g. white piece of paper or wall). One will then perceive, for a few seconds, a vague 'negative' version of the target picture. Dark regions of the original have become light and vice versa, and chromatic colours have changed into their complementaries, i.e. red becomes green, blue becomes yellow, etc. (Plate 3.2). The negative afterimages are highly interesting with regard to questions concerning the physiological mechanisms for colour vision, and they may be explained as being evoked by 'exhaustion phenomena' in the eye (for further information, see Section 4.5.3). Furthermore, the negative afterimages provide a direct demonstration of the concept and existence of complementary colours.

Plate 3.2 The appearance of complementary colours in negative afterimages (example inspired by the Swedish background of the author). Fixate the midpoint of the blue cross and don't allow your eyes to wander elsewhere. Then, after at least 20 sec of staring, quickly move your eyes to a white sheet of paper. You might then see a very weak yellow cross with a blue background. After a few seconds, this afterimage fades away. A black and white version of this figure will appear in some formats. For the colour version, please refer to the plate section.

Besides the negative afterimages there are also more shortlived and less complex *positive afterimages*, which have the same colouring and lightness patterns as the original target. These afterimages appear after looking at the target for only a few seconds and then closing the eyes or switching off the light. One may then notice that the original visual perception does not disappear immediately, but it becomes more vague and gradually wanes in the course of, roughly, a second. In various ways, this kind of afterimage is very useful for us. Thanks to the positive afterimages our visual perception does not get interrupted each time we blink. Furthermore, thanks to these afterimages we can use common types of film and video techniques: a sufficiently rapid series of still pictures gives us the illusion of a continuous view and if the consecutive still pictures change in an appropriate manner we see movements. In movies, the intervals between successive pictures should be about 1/30 s or less in order to make them merge for our eyes and brains. Thanks to mechanisms associated with the positive afterimages, we will also get the illusion of a steady light from a flickering one if the rate of ON/OFF switching is rapid enough. The minimum rate for this *flicker fusion frequency* is highly dependent on light intensity and on the size of the illuminated area (for further information, see Section 4.5.3).

3.1.3 Colour constancy and coloured shadows

It is well known by most people that the perceived colour of an object may change with illumination: e.g. as seen in sunlight versus in the light of incandescent lamps or candles (cf. *metamerism*, Section C.4). However, in spite of the fact that we live in surroundings with very variable kinds of illumination, we are usually not aware of correspondingly large alterations in the hues of the reflected colours. Objects look rather similar under the midday sun and at sunset, and a white paper looks white both in broad daylight, at dusk, and under artificial light indoors. This perceptional phenomenon is called *colour constancy*, and it depends on highly complex calculations and corrections made continuously by our brains, adjusting the colour perceptions such that objects remain recognizable in spite of the illumination-dependent changes in their reflected light. It is biologically important to be able to identify things in our surroundings and find adequate food at all times, both during daytime weather variations and at different levels of evening dusk.

The brain's settings for colour constancy depend, from moment to moment, on the total visual environment.[73] For instance, while incandescent lamps were still widely used, windows of lighted rooms looked yellow when observed from the street at night. When being inside such a room, however, the brain uses a different setting for colour constancy, and the lighting is then essentially seen as white.

One of the by-products of colour constancy is the phenomenon of *coloured shadows*, which may appear if a white and a coloured source of illumination are used simultaneously and from different angles. In the case of Plate 3.3, sunlight and a lamp emitting blue light are both illuminating a vertical candlestick. The background is a greyish wall, on which the candlestick throws a shadow with a separate localization for each source of illumination. In the sunlight-blocking shadow, the lamp provides the only remaining illumination and, as expected, this shadow is blue. In the lamp-light-blocking shadow, the only remaining illumination is the sunlight, and this shadow might be expected to be grey. However, for a human observer this shadow does not look grey but yellow (Plate 3.3a, b, 's'), i.e. it is coloured with a hue complementary to the blue illumination. The hue of this coloured shadow is fabricated by the brain of the observer, and its colour is dependent on the composition of the whole scene in front of the eyes. At the original scene, this may be checked by looking at only the isolated wall area of the coloured shadow, using a simple viewing tube of rolled-up black paper. Looked at in this way, the coloured shadow loses its colour and the wall indeed looks grey (cf. 's' in Plate 3.3c, d). Paradoxically, the illusion of coloured shadows may be photographed (Plate 3.3a, b) in spite of the fact that, for a human observer present at the scene, it depends on corrective brain mechanisms. This is possible because digital cameras are programmed to imitate corresponding aspects of human vision: also the camera makes

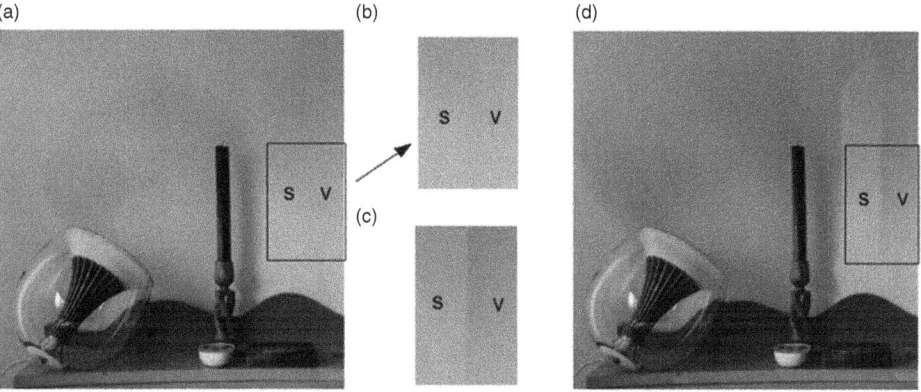

Plate 3.3 Demonstration of the complex phenomenon of *coloured shadows*. An unlit candle-stick is standing on a cupboard, being illuminated both by the white sunlight from a left-side window and by a lamp with blue light. For each source of illumination, the candle-stick throws a shadow on the greyish wall. Panel **(a)** shows the scene as it is perceived by a human observer: the left shadow is dark blue and the right one (s) is weakly yellow. In panel **(d)**, the photographic colour-settings have been adjusted, and shadow s now has the greyish colour it should have if simply showing the light reflected from this part of the wall. In order to facilitate colour comparisons between the two versions of shadow-s and nearby portions of the wall (v), panels **(b)** and **(c)** display these details separately, as taken from panels **(a)** and **(d)**. Further explanation in Text and below. A black and white version of this figure will appear in some formats. For the colour version, please refer to the plate section.

Panels (a) and (d) are from exactly the same photograph, taken in a so-called RAW format. For panel (a), the white balance had been automatically handled by the camera, trying to imitate the colour constancy of a human observer. Panel (a) does indeed closely correspond to how the original scene looked for someone with normal colour vision. Afterwards, in a separate copy of the digital photograph (panel (d)), the white balance was re-adjusted using photographic software. With such computer programs, the white balance may be re-calculated for a selected target region within the image. The RGB mixtures for all the colours of the whole image are then adjusted such as to make the selected target region (almost) achromatic. For panel (d), the white balance was set for a target region in the lower-left corner of the image, a site that was not illuminated by the blue lamp. This setting would have been adequate if the scene had only been illuminated by sunlight from the window.

corrective calculations for the various colours within the scenery, taking the illumination into account, before arriving at its final displayed picture.[74]

In digital cameras, colours are corrected for effects of illumination using internal calculations for the so-called 'white balance' (Section D.3). In the photographic picture ultimately shown by the camera on its display, the colours have ideally been adjusted such that a white paper continues to look white. When photographing a blue-lit scene like that of Plate 3.3a, the camera will automatically adjust its white balance such that at least part of the wall continues to look grey (Plate 3.3a, b, 'v'). This partial neutralization of the blue illumination produces the side-effect that some grey areas outside the blue illumination tend to become yellowish; in panel A this concerns the shadow 's' and regions to the left, below the lamp. If the same scene is photographed with the white balance specifically set to be valid for regions illuminated only by the daylight (this can easily be done with modern cameras), then a larger piece of the wall will be shown as blue and the coloured shadow will lose its colour (cf. Plate 3.3c, d, white balance here set for the lower leftside corner).

When looking at the real world, without the assistance of a camera, coloured shadows of scenes like Plate 3.3a are largely the by-products of the white-balance calculations of our brains.[75] In addition, mechanisms for simultaneous contrast might play a role. In Plate 3.3d, the blue colour surrounding the lamp-light-blocking shadow might give a sensitive observer the impression that this shadow still contains traces of yellow. Such effects disappear if this portion of the picture is isolated from its blue surrounding (Plate 3.3c).

The coloured shadows and phenomena concerning colour contrast have long fascinated various scientists. For instance, in Paris in 1789, just a few weeks prior to the onset of the French Revolution, the French mathematician Gaspard Monge (1746–1818) gave a surprisingly modern lecture on colour constancy.[76] More recent demonstrations of the brain's remarkable calculations and corrections for colour constancy have been given by Edwin Land (1909–91), an American physicist and vision-system physiologist who also co-founded the successful Polaroid company for the fabrication of instant cameras and polarizing sunglasses. In the 1950s, he experimented with colour pictures that were shown in the classical manner using three projectors: one with a red filter, one with a green filter and one with a blue filter. The pictures shown had all been photographed with cameras provided with the corresponding filters and mono-chromatic film (cf. Maxwell's colour photograph from 1861, Plate 2.5c). At the end of such a projection session, Land switched off the blue projector and removed the filter from the green one while it was still on. He then happened to look at the screen: all the colours still seemed to be present! Land as well as his assistant could see blue, green and yellow colours in a projected picture that should have shown only a mixture of red and black/grey/white.[77] These observations led to a series of new experiments in which Land demonstrated the unexpectedly strong influence that the picture as a whole has on the apparent colours of its various details, all of which was also highly dependent on the wavelength composition of the illumination.

Land developed a model of his own for how colour constancy works, and he called it the *Retinex theory*, a name implying combined effects of the retina of the eye and the cortex of the brain.[78] However, neither the different mathematical versions of Land's Retinex model, nor various other competing and later models, have so far been able to completely explain the complex phenomenon of colour constancy or the interesting effects of two-colour projections. There has been much recent research in this field, concerning mechanisms and conditions for colour constancy as well as regarding the apparent colour interactions within abstract and complex scenes.[79] In situations imitating natural changes of illumination, colour constancy is generally good but not perfect; object colours may look approximately but not precisely the same under different light sources. The sensitivity to illumination changes is particularly high for achromatic colours, e.g. for white papers staying white (cf. Section C.4). Within a natural and complex scenery under varying illumination, the colour constancy of an object might become disturbed by certain manipulations in its surroundings, e.g. by changes in the reflectance of neighbouring objects. Furthermore, the constancy is less good if the change of illumination fails to alter the average light intensity or that of the most intense visual stimulus within the field of view. Correspondingly, the brain's estimate of an illumination change may possibly be based on measurements (receptor responses) of how the average light intensity for different wavelength bands is altered across a whole scene, or how it is changed in the brightest patch in view. There are many other cues of possible importance, such as specular reflections from surfaces, and various shadows and mutual inter-reflections among objects. Furthermore, prior knowledge about how the illuminated objects look in the natural world might also be of importance.

It is still a matter for further analysis and discussion what the relative roles are, for these adaptive processes, of first-line mechanisms in the eye (e.g. sensory cells) versus those at higher levels in the brain (see Section 4.7). Some of our basal colour perceptions are themselves dependent for their very existence on mechanisms for simultaneous contrast and colour constancy. Among commonly named colours this is, in particular, valid for brown, grey and black. Brown is generally described as a dark and/or unsaturated version of yellow, orange or red. However, light of the same wavelength composition might look brown if the background is relatively dark, but yellow if the surroundings are perceived as being light (cf. Plate 3.1). Black is usually described as being synonymous with darkness, the absence of light. However, our perception that an object looks black depends very much on the simultaneous contrast between the object and its surroundings and on the general illumination of the scene. Persons remaining for some time in complete darkness usually report seeing some kind of homogeneous grey ('brain grey') rather than deep black. In a projected film, various details might look deep black in spite of the fact that no part of the screen reflects less light than the greyish one seen in the absence of projection. However, grey colours are typically seen only in a context of contrasting luminance (apparent exception: the brain grey mentioned above).

3.2 Alternative colour theories: Goethe and Hering

As was described in the preceding chapter, a revolutionary progress in our understanding of colour vision was generated by Newton's prismatic experiments and the subsequent empirical insights concerning trichromatic colour perception (Young, Maxwell, Helmholtz, etc.). However, as is commonly the case at times of scientific paradigm shifts, there was also much doubt and criticism. Two of the best-known opponents of the new trichromatic colour theories were Goethe and Hering. Much of the criticism concerned perceptual phenomena which were not easily explained by simply assuming a fixed relationship between the wavelengths of the light and the sensitivities of three (hypothetical) colour sensors. As was described above (Section 3.1), small fields of vision often show an evident relationship between wavelengths and colour, whereas for larger fields, containing different colours, these relationships become more complex, due to the influence of various aspects of cerebral colour processing.

3.2.1 Goethe and his *Farbenlehre*

One of the early investigators in colour psychology was the world-famous literary author Johann Wolfgang von Goethe (Plate 3.4a). Over many years, he performed perception-psychological studies of colour vision, and his final account of all these efforts was published in 1810 under the title, *Zur Farbenlehre* (about 2000 pages). Goethe considered this work his most important achievement and publication. His book had three main parts, and the second one had the polemic title *Enthüllung der Theorie Newtons in welchem er versucht Newtons Farbtheorie zu widerlegen.*[80] For instance, Goethe refused to accept that white light was not homogeneous and, without having any scientific proof, he thought that colours appeared due to interactions between 'light' and 'darkness', an idea similar to that of philosophers of the ancient world. He did not accept that darkness may be understood as an absence of light.

In a relatively modern manner, Goethe classified the colours into three categories:

1. *Chemical*, depending on the reflection and absorption of light in various materials (cf. absorption colours, Section 2.8.1).
2. *Physical*, depending on the structure of the coloured object (cf. structural colours, Section 2.8.2);
3. *Physiological*, depending on physiological and perception-psychological properties of the observer.

From a modern point of view, Goethe's most important contributions concerned category (3), including observations that could not easily be explained by the findings of Newton. However, on the whole, this concerned various subjective colour phenomena that Newton did not study. Newton himself was conscious of such general limitations; in his book *Opticks* (1730) he declares that: 'I speak here of colours so far as they arise from light. For they appear sometimes by other causes.'[81] The subjective experiences studied by Goethe included the 'coloured shadows', which we now know have to do with the complex subject of colour constancy (Section 3.1.3, Plate 3.3). Goethe was also fascinated by the phenomenon of negative afterimages (Section 3.1.2), and he even painted a portrait of a woman in complementary colours, made such that it would look 'normal' only in its negative afterimage. Inspired by his experiences with coloured shadows and negative afterimages, Goethe organized the various colours in pairs of opponents: red versus green, yellow versus blue; in this respect he was a predecessor of Hering (Section 3.2.2).

Goethe developed systems of his own for arranging collections of colours in circles and triangles, and he believed that each colour had its own psychological effects (Plate 3.4b). Together with an author colleague, Friedrich Schiller, he published *Temperamentrose* (1799), a circular diagram in which specific colours were coupled to each one of the four classical temperaments and their associated traits of character and personality. Goethe's many observations and ideas on colour have served as a source of inspiration for numerous famous artists (e.g. Turner, Kandinsky, Klee, German expressionists), and his colour theories are still emphatically cited in many non-scientific contexts. In particular, he has had (and has) an influence on various anthroposophic and other mental systems in which colours are directly associated with various esoteric moral and/or emotional values (cf. Section 1.5).

(a)

Plate 3.4 (a) *Johann Wolfgang von Goethe* (1749–1832), portrait painted in 1828 by Joseph Karl Stieler. As part of his extensive colour studies, Goethe analyzed phenomena that are not easily explained knowing only the spectral light composition (e.g. afterimages, coloured shadows).
© Peter Horree/Alamy.

(b)

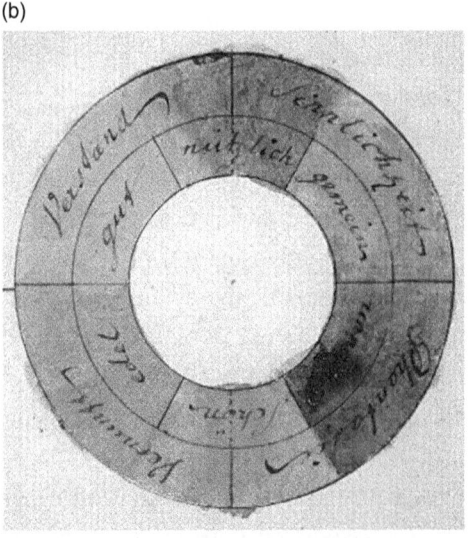

(b) In his colour charts Goethe arranged the colours in various ways, e.g. in a circle showing colours together with their (assumed) psychological effects.
Source: http://en.wikipedia.org/wiki/File:GoetheFarbkreis.jpg.
A black and white version of this figure will appear in some formats. For the colour version, please refer to the plate section.

3.2.2 Hering and his theory of colour opponents

One of the major antagonists to the trichromatic Young–Helmholtz theory was the German physiologist K. E. K. Hering (Plate 3.5a), a professor in Prague, who launched his *Opponent Colour Theory* in 1892. His perception-psychological and introspective studies led him to the conclusion that there were six unique 'elementary colours' (in German called *Urfarben*) that could neither be made nor described as the result of ('mental') colour mixing. Hering's four chromatic opponents included red versus green and yellow versus blue; in addition, his pairs of *Urfarben* included white versus black. There are no red-green or blue-yellow transitional colours; for each one of these pairs, additive mixtures may become achromatic, greyish/white. Hering also investigated the behaviour of complementary colours in negative after-images, and all his various studies led him to the (false) conclusion that there are not three but six different kinds of wavelength-specific light receptors in the eye.

Hering arranged collections of colours in a circle with the chromatic opponents in front of each other (Plate 3.5b), rather similar to colour circles of Goethe (Plate 3.4b). With regard to general principles, both kinds of circles are, of course, inheritors of Newton's original colour circle (Plate 2.1c). Between each one of the four chromatic 'unitary' (= primary or elementary) colours, Hering's system included all the other hues as gradual transitions of 'binary' (= secondary) colours. Hering's theories and colour models have had a marked influence on various modern systems for technical colour ordering and encoding.[82]

Inspired by Hering's investigations, many studies have been performed in order to find out how well individual subjects agree in their identification of each one of the four 'unique' chromatic opponents.[83] In psychophysical experiments, the participant is asked to make a choice of introspectively unique hues from, for instance, the continuously varying hues of a sunlight spectrum. 'Find a blue colour with no trace of yellow or red', 'find a green without blue or yellow', and so on. In such investigations, each person will generally indicate almost the same hue on different occasions. Typical wavelengths for unique elementary hues are about 475 nm for blue, 500 nm for green, 580 nm for yellow and $\geq$650 nm for red. Red is a special problem in this context because it is never felt to be truly unique; spectral colours are typically seen as being very similar in hue between 650 and >700 nm, and they are all perceived as red with traces of yellow.

(a)

Plate 3.5 (a) *Ewald Hering* (1834–1918) developed an "*opponent colour system*", centred around two pairs of chromatic and one of achromatic elementary colours ("*Urfarben*").
Image courtesy of The National Library of Medicine.

(b)

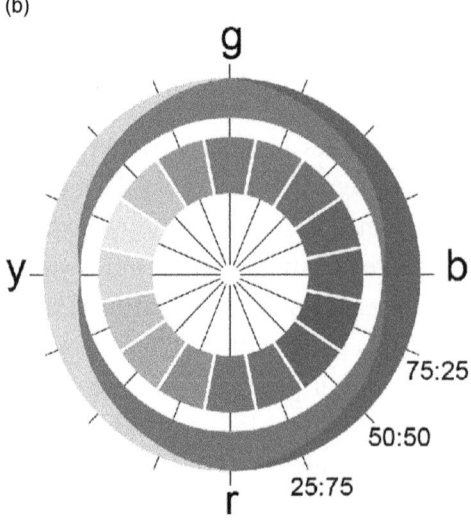

(b) Hering's colour circle with chromatic opponents along the periphery (green/red and yellow/blue) and white/black in the middle.
Source: http://commons.wikimedia.org/wiki/File:Ewald_hering_colors.jpg
A black and white version of this figure will appear in some formats. For the colour version, please refer to the plate section.

In comparisons between different individual participants, 'unique' spectral hues may show a considerable variability. In one investigation it was even observed that wavelengths considered by some to be 'uniquely green' overlapped with wavelengths seen by others as 'uniquely blue'.[84] It is understandable that even people with normal colour vision may sometimes disagree on precisely which colours they are seeing in their surroundings.

With increasing light intensity, most spectral wavelengths will show minor changes in hue; e.g. yellow-green colours become more yellow.[85] According to some investigators, the unique primary hues of blue, green and yellow differ from all the other spectral hues by not being intensity sensitive;[86] however, in some laboratories such an enhanced stability of unique hues was not observed.[87]

Hering's conviction that, for instance, yellow is a unique colour may be consistent with the introspective feelings of many participants: seeing a yellow surface, it might be difficult to imagine that this colour could be produced by mixing other hues. However, physically this is not correct: in additive mixtures, yellow is easily produced by combining red and an appropriate non-opponent version of green (e.g. c. 546 nm; Plate 2.11). Furthermore, natural yellow colours are commonly the result of a light mixture containing very many longish wavelengths (Plates 2.11, 2.13b). It is interesting to note that, during earlier periods of art history, yellow and green were often considered as variants of the same colour;[88] it is evidently not obvious to all human subjects that yellow and green are two completely separate and unique primary colours.

In this context it is important to remember that, even in a modern industrialized society, each one of the few basal colour terms of common everyday language (cf. Plate 1.2) might cover spectral hues with quite varying properties. For instance, the term 'green' is generally considered to cover the range of about 490–560 nm. Within this range, light of the shorter wavelengths (e.g. 495 nm) would combine with red to give white (i.e. be complementary to red), while light of longer 'green' wavelengths (e.g. 546 nm) would combine with red to give yellow (for further information, see Section B.1.2 and Plate B.2).

Interestingly, modern physiological investigations have led to a peaceful resolution of the apparent conflict between the Hering and the Young–Helmholtz theories: they are both correctly summarizing important aspects of how colour vision takes place. The receptive mechanisms of the retina are clearly trichromatic, in good accordance with the Young–Helmholtz model. The subsequent processing of the colour information within the retina and brain is, however, largely organized in a manner that seems to fit well to the Hering model (see Chapter 4).

3.3 Our strange visual field, our eye movements and the creativity of our brain

The creative contributions of our brain to vision are not restricted to the phenomena of colour contrast and colour constancy. There are also many other intriguing differences between the physical properties of the picture projected into the eye and the resulting perception produced by the brain. For a further discussion of these things it is important first to consider the hardware organization of our visual field.

When investigating what is perceived through an eye in its different viewing directions, the owner of the eye is typically seated in a chair with a support for the head and he/she is asked to fixate on a point straight ahead (e.g. a light or cross). While the participant is gazing steadily at the target, small light flashes or letters may be shown at different positions within the visual field. Through the lens system of the eye, an image of the outside world is projected onto the back side of the eye, which is covered by a light-sensitive tissue (*retina*). Each point within the visual field corresponds very precisely to a projection-point on the retina. As compared to the outer world, the retinal projection is upside-down and left/right-mirrored. With regard to the position and size of various things within the visual field, it is impractical to use absolute measures like centimetres or metres, because the relative sizes and positions of objects also depend on their distances from the eye. Therefore, in this context, the most appropriate method of quantification is to use angular measures, horizontally and vertically. Straight ahead the visual angle is $0°$, straight to the left or right the horizontal visual angle is $90°$, etc. Similarly, the size of an object may be given by its subtending visual angle; e.g. a full moon subtends an angle of about $0.5°$.

As everybody knows, our total field of view is very wide (Figure 3.1). Each eye can see further out towards the side than towards the middle because, in the middle, the nose blocks the view. Horizontally, each eye may look about $100°$ toward the side and about $60°$ toward the middle. This means that the total horizontal field of vision is about $200°$, out of which about $120°$ may be seen simultaneously by both eyes (Figure 3.1). Using two eyes is not only important for a wide total field of vision, but being able to use both eyes simultaneously is also important for our ability to perceive how far away things are from the eyes, our binocular stereoscopic three-dimensional vision. Using our eye muscles, we are continuously (re-)directing our eyes toward various objects within our visual field and, with enormous precision, this is normally automatically done for both eyes together such that they are always both looking at the same things (Figure 3.1). This requires that the directions of fixation of the two eyes must co-vary in different ways for nearby and for more remote objects: the fixation angles must converge more for close than for distant things. This phenomenon is easily observed: ask a willing person to gaze at your finger in front of his/her face. Gradually bring the finger closer to the face; you will then see how the two eyes will gradually become directed further and further toward each other (convergence).

3.3.1 All eyes are halfblind

As we know, but rarely think about, we only have a sharp vision across a very small region in the middle of our field of view. When reading, this sharp region often covers only a few words. At all sides around this minor sharp region, the visual acuity decreases very rapidly (Figure 3.2). Already at a distance of c. 2.5° from the centre, the acuity has gone down by more than half, and further peripherally things become still worse. Apart from the minor central region, corresponding to the *fovea* in the retina, we are actually halfblind. In spite of this, we may still have the feeling that we see everything clearly and sharply focussed across the whole visual field. This agreeable illusion comes about because we are systematically and rapidly directing the sharp region of our eyes toward all details that we judge to be interesting (Figure 3.3b). Continuously and rapidly, the brain produces a representation of a whole picture that we believe we have seen, based on a combination of all the small sharp fragments plus the larger unsharp views from our peripheral vision (cf. Figure 3.3).

The wide peripheral regions within our visual field are used for rough perceptions that may cause us to direct our eyes there and discover important details or, for instance, threatening dangers. In this context it is essential that our ability to detect movements is particularly great for the peripheral visual regions.

3.3.2 Eyes must move to see

Each eye is moved and stabilized by six different *extrinsic eye muscles*, which belong to the most rapidly contracting muscles in the body. When we direct our eyes toward whatever we wish to look at in detail, we use fast types of eye

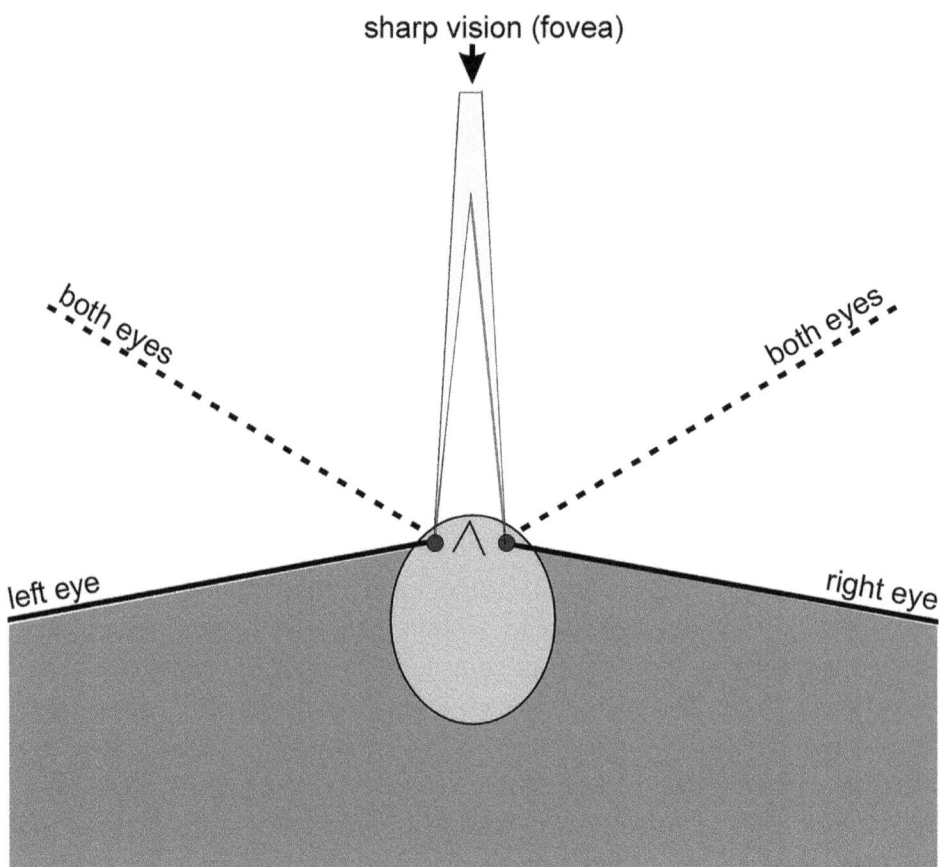

Figure 3.1 Scheme illustrating the difference, in humans, between the wide total field of vision and the very much smaller field for sharp vision (fovea).

movements called *saccades* or *saccadic* eye movements. During the rapid movement itself, our reception of visual signals is suppressed such that we will not be disturbed by a blurred picture. After the movement, the eyes are held relatively still during a fixation period of a few tenths of a second (cf. the exposure time in a camera), used for collecting visual information from this point in the visual field. Thereafter, the eyes are rapidly moved to a new position, collect data from a new point of fixation, etc. On average, we perform several saccades per second during the whole of our daily waking period.

Besides the saccades, there are three other kinds of automatic eye movements.

– *Vergence movements* (convergence, divergence; already mentioned above), which are needed to enable the two eyes to look simultaneously at the same object, whether close-by or far away.

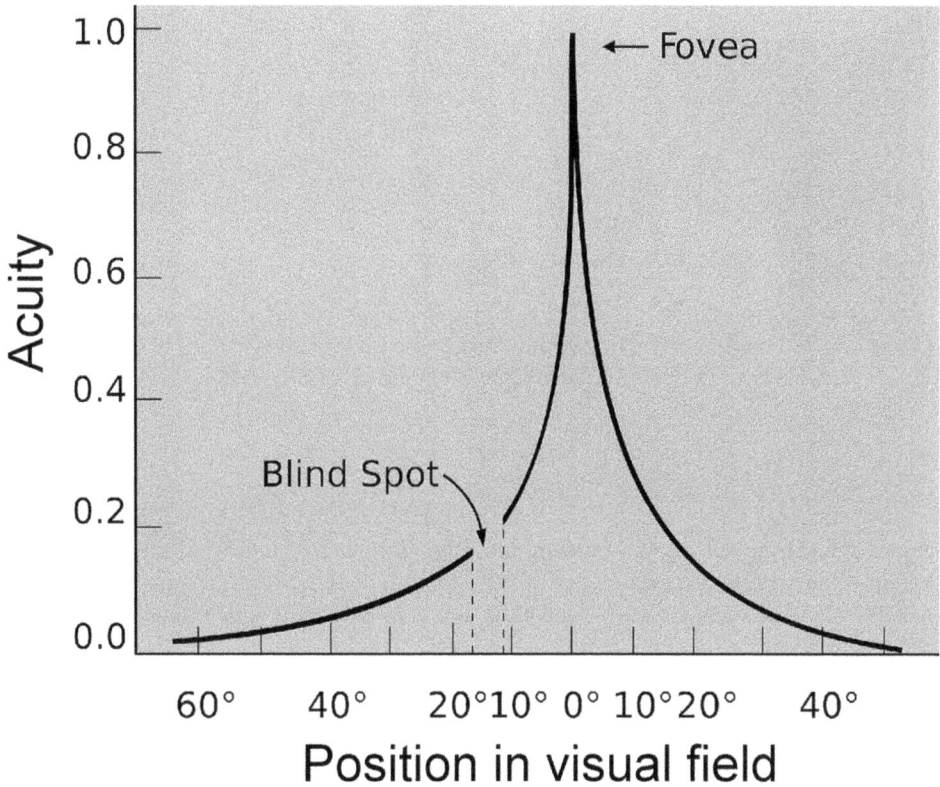

Figure 3.2 Visual acuity within different parts of the visual field, as determined for a left side eye. The region with an acuity of >0.5 is about as large as the blind spot. (After Wertheim, 1894.).
Source: http://en.wikipedia.org/wiki/File:AcuityHumanEye.svg.
Image by Vanessa Ezekowitz.

Both of the other kinds of automatic eye movements serve purposes of image stabilization:

- *Optokinetic reflex movements*, including a *smooth pursuit* phase which enables the eyes to keep fixating on a moving object. Thanks to this reflex one might, for instance, easily follow a moving child or car with the eyes, keeping the object within the sharp region of the visual field.
- *Vestibular reflex movements*, which help to keep the eyes fixated on an object during rapid movements of the head. Together, the latter two stabilization reflexes are effective enough to make it possible to read a text while rapidly shaking the head, something which even a modern digital camera would find difficult.

(a)

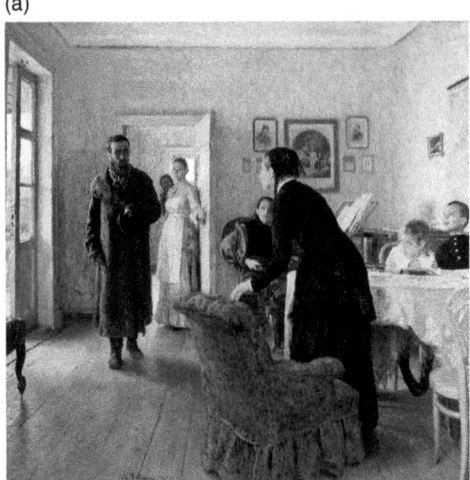

Figure 3.3 (a) A well-known Russian painting, "*Unexpected return*" (1884) by Ilya Repin. While people were looking at a reproduction of this painting, their eye movements were recorded and it was thus determined toward which details their eyes became directed.
© Heritage Image Partnership Ltd./Alamy.

(b)

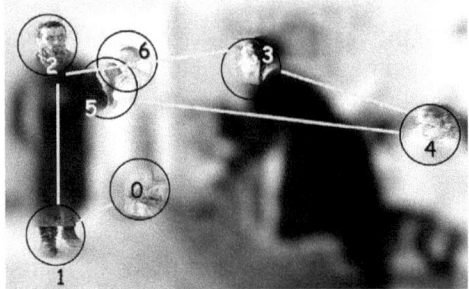

(b) shows, for one subject, the recorded sites of eye fixation dring the initial 2 seconds of looking at the painting: the first 6 eye positions included 4 faces.
Source: http://en.wikipedia.org/wiki/File:Vision_2_secondes.jpg.
Image by Hans-Werner Hunziker (after Yarbus, 1967).

The normal stabilization of the direction of gaze is not perfect, and was not meant to be. Even during steady fixation, in the absence of any evident changes in the direction of viewing, the eye is still trembling a little. Two types of such spontaneous and minute movements occur simultaneously: *micro-saccades* and *ocular micro-tremor*. These various 'micro-movements' do not, however, make the picture more unclear but rather the opposite:

1. the brain has internal mechanisms compensating for the minor blurring caused by micro-movements;
2. in the complete absence of continuous micro-movements the eyes are unable to see anything at all.

This latter point may be investigated using a minute projector placed on the eye itself, shining a picture onto the retina. This projector follows all movements of the eye and may thereby produce a completely stabilized picture on the retina. After its onset, the perception of such a picture rapidly becomes more vague and it disappears altogether within less than half a minute (perhaps after a few seconds). This is probably due to a property that visual sensors and neurones have in common with the cells of many other sensory systems: they are typically much more sensitive to signal changes than to the absolute intensity level of the signal. Consider, for instance, the sense of touch in your fingertips. In order to feel the surface structure of a piece of textile or paper, you have to move the fingers across the target surface. While keeping the fingers still, nothing much is felt at all.

The invisibility of stabilized images has at least one functional advantage: it helps to save us from seeing the shadows of the many rather large blood vessels that are running across our retina. For similar reasons, we do not normally see the rather large speck of yellow pigment which is covering the fovea and adjacent parts of the retina (yellow spot, *macula lutea*). However, in addition to the fading of stabilized images, the online 'retouching' mechanisms of the brain are of great importance for covering up any holes or structural imperfections in the visual input (see discussion of the blind spot, Section 3.3.4).

3.3.3 All eyes are partly colour blind

The colour sensors of the eye (the cones, Figures 4.2 and 4.3) are not uniformly distributed within the retina, and this is one of the major reasons for regional variations of colour perception across the visual field. The normal trichromatic colour vision is limited to a relatively small area at the centre of the field. Detailed measurements of human colour vision have usually been done for central regions of $2°$ or $10°$ divergence. Also within such small retinal areas there are, however, measurable variations in colour vision, and differences become more drastic further out in the periphery. Below, a brief summary is given of these observations, all concerning subjects with normal trichromatic colour vision:

1. Normal eyes are 'blue-blind' within a small central region of 0.35° divergence; within this retinal portion there are no strongly blue-sensitive sensors (no S cones, see Section 4.3.3).
2. Normal eyes are dichromatic within a rather extensive peripheral region of the visual field, at about 25–30° or more from the centre.
3. Normal eyes are completely colour blind in the most peripheral regions of the visual field, at about 40–50° or more from the centre.

Measurements concerning points (2) and (3) show some variation in relation to the state of dark adaptation of the eyes. Furthermore, colour vision is generally better preserved in regions toward the side than in those toward the midline.[89] However, it is quite clear that colour vision is very deficient within large peripheral areas of the visual field. Normally, these deficiencies will pass unnoticed because the images that reach our consciousness are made by the brain and include 'correct' colours for all interesting items, using information as 'seen' by the central trichromatic regions thanks to appropriately directed saccadic eye movements (cf. Figure 3.3b). The central 'blue-blindness' (1) is normally also very effectively concealed by the brain; if regions near the centre of the visual field are bluish, the brain also applies a suitable shade of blue to the most central portion (cf. next section).

3.3.4 All eyes have a hole in the visual field

The retina has light-sensitive cells everywhere except at the joint site of exit for all the fibres of the optic nerve (Figure 4.1, *optic disc, papilla*). Hence, that portion of the retina is blind, and this 'blind spot' is even larger than the fovea, the region of sharpest vision (Figure 3.2). The optic disc has a size of about 1.86 × 1.75 mm. It replaces the retina within a region covering about 6–6.5° of the visual field. Such a gap in the field of vision is not insignificantly small: its width corresponds to at least 12 full moons placed beside each other.

In spite of its considerable size, the presence of the blind spot is normally not noticed at all, not even when looking at the surroundings using only one eye. This is a remarkable feat, and it illustrates how well the brain performs its many sequential corrections and adaptations of visual images before they reach our consciousness. We never see the original signals generated in the retinal sensors, we only see what the brain shows us, after its extensive interpretations and adaptations of the image.

X

X

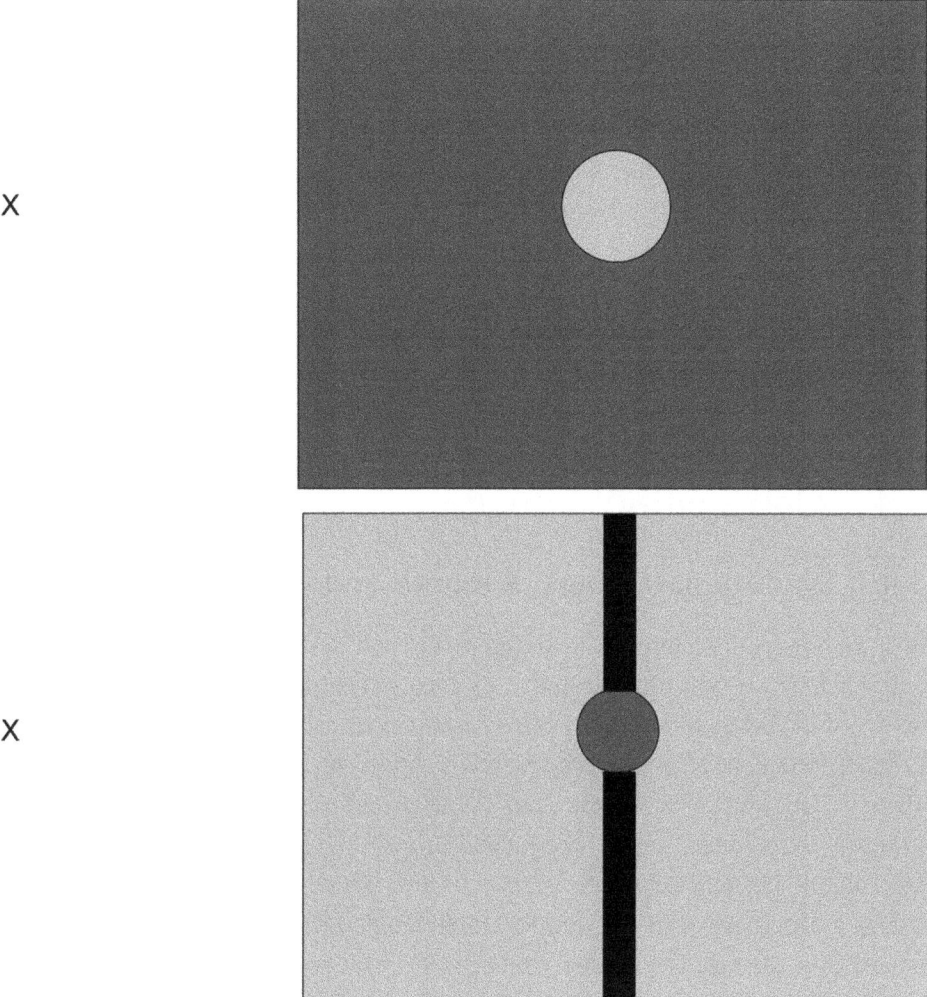

Plate 3.6 Charts that may be used for demonstrating how the brain continuously retouches the blind spot, providing it with colours and patterns from the immediate surrounding. *Instruction for use*: see Text. A black and white version of this figure will appear in some formats. For the colour version, please refer to the plate section.

With regard to the blind spot, the brain's sequential corrections happen such that the 'hole' becomes filled up with the colour and coarse visual pattern (if any) of its immediate surrounding. The outcome of such processes may be experienced using the two panels of Plate 3.6. Both panels were made to be seen with the right eye; keep the left eye covered. Hold the upper panel of Plate 3.6 at about 40 cm in front of the eye, and stare steadfastly at the little cross (upper left side); do not allow the eye to wander elsewhere. Slowly bring the page closer to the eye. At a certain distance, commonly at about 20–30 cm, the yellow circle will suddenly disappear and the whole of the surface becomes blue. Slowly bring the book still closer to the eye; at a certain distance the yellow circle will suddenly reappear. The lower panel of Plate 3.6 shows, correspondingly, that the blue circle disappears. Furthermore, in this case the interrupted thick black line will become 'repaired': when the blue circle disappears, the black line will seem to be going continuously from top to bottom. The diameters of the circles in Plate 3.6 demonstrate the considerable size of the blind spot (even somewhat larger circles might have been used with similar results).

3.4 Colours without images

3.4.1 Synesthesia: colours of sounds and letters

In a dictionary, the term *synesthesia* is defined as 'a concomitant sensation; esp. a subjective sensation or image of a sense (as of color) other than the one (as of sound) being stimulated'.[90] To a certain extent this kind of sensory analogy is quite normal, as is apparent in everyday language. For instance, one may talk about a 'dark tone' or a 'warm colour'. With regard to spectral hues, their relative 'warmth' or 'coldness' tend to be judged in a similar way for different subjects: red colours are generally seen as 'warm' and bluish colours as 'cold', perhaps due to associations with fire and ice. Interestingly, these connotations are quite opposite to the relative energy contents of the respective photons: shortwave blue light is more energetic than longwave red light (Section 2.4).

For some individuals, the feeling of analogies between different sensory modalities is drastically enhanced and forceful: certain abstract concepts or sensory stimuli might automatically generate a strong kind of secondary sensory 'perception'. For such subjects, individual numbers and letters might all have their own separate colour, a moving or flickering light might become accompanied by a sound, the days of the week and various names of persons might each be associated with a specific colour, etc. Many different variants of such a 'neuro-synesthesia' have been described[91] and it has been estimated that such traits are present for about 1% or more of the Western population. Synesthetic links to colour are very common, being found for 80% or more of all reported cases of neuro-synesthetic experiences.[92] For instance, links from sound to colour are often present: listening to music might give the subject the simultaneous experience of seeing vivid colours. The synesthetic colour experience may be associated with 'secondary perceptions' related to other surface features, e.g. the feelings of touching metal, velvet, etc. Links between primary triggers and secondary perceptions are not consistent between different neuro-synesthetic subjects: for one person the number 4 might be blue, for another green or yellow. There are several instances known of artists with more or less evident neuro-synesthetic properties.[93] Commonly cited examples include the painter Kandinsky, who was a pioneer in non-figurative art and found this analogous to music, and the poet Rimbaud, who described in a poem the colours of French vowels.

3.4.2 Other examples of colours without images

Besides the synesthetic states and other items mentioned earlier in this chapter, there are still other examples of how colour experiences might be generated by eyes and brains in the absence of any retinal analysis of wavelength differences. Thus light (flashes) and colours may be seen when:

1. pressure is applied to an eye;
2. cosmic radiation hits an eye (primarily concerns astronauts);
3. electrical stimulation is applied to an eye or to visual regions within the central nervous system;
4. there is an imminent attack of migraine: perceptions of visual patterns or flashes, sometimes coloured, often precede the intense migraine headache by about 60 min or less (aura); sometimes such a state occurs without the succeeding headache;
5. hallucinations occur, as caused by hallucinogenic drugs (e.g. LSD), brain damage or disease;
6. drugs or brain damage cause *chromatopsia*, a state in which everything seems tinged with a particular colour (Section 6.3).

Visual perceptions which are not dependent on light entering the eye are called *phosphenes*.

There are also visual illusions in which colours unexpectedly appear out of colourless surroundings. One classical example is *Benham's top*, first made in 1895 in the toy factory of Benham. The scientists Fechner and Helmholtz had already earlier observed the phenomena concerned. Benham's top has a circular upper side, half is black and the remainder shows a black-and-white pattern of circular lines. When the top is rotated at sufficient speed, relatively low-saturated colours will appear along some of the concentric circular lines. Different persons might see different colours, and the coloured lines might change their position if the direction of rotation is reversed. This coloured illusion probably depends on differences in speed between the various kinds of retinal cones and their respective neural networks. An image of a rotating Benham's top may be seen on the internet.[94]

4 Our biological hardware: eye and brain

In this chapter I will give a very brief summary of how our biological 'vision machinery' is constructed and how it works, particularly with regard to colour vision. The eye and the brain belong to the most complex organs in the body and the whole book might easily have been filled with only these subjects.

4.1 Our three main kinds of vision

Traditionally, humans are considered to have five senses: sight, hearing, touch, smell and taste (more modalities might be added, like pain, equilibrium, etc.). All the traditional five are essential for a normal life, but the one that the majority of people would be most unwilling to lose is probably sight; this, in spite of the fact that a blind person probably becomes less socially isolated than somebody who is completely deaf. Compared to many other species of mammals, humans, apes and monkeys are unusually dependent on vision in their normal behaviour and interactions. From an evolutionary point of view this is understandable: a good sense of vision is essential for climbing and moving around in the trees of jungles and savannahs. Consequently, a large proportion of our brain is devoted to various kinds of visual analysis. What are we seeing and how? What is the relationship between colour vision and our other visual capacities? How important are the colours?

The great importance of our visual capacities is further underlined by the fact that we may be said to have three different kinds of vision, two for use in daylight and a third one for darkness:

1. Our *luminosity vision* (vision of achromatic *lightness contrast*, i.e. 'black-and-white' vision). In normal daylight, our eyes and brain are extremely well equipped to discover the edges of contrast between darker and lighter regions in the visual field. This capacity is the most essential one for our visual orientation and for recognizing structures and objects in the surroundings. In the central portion of the visual field, such functions may concern very small details (Figure 3.2).
2. Our *colour vision*. This kind of vision is also used in normal daylight, and we are well equipped for discovering differences in colour and edges of colour contrast.
3. Our *night vision*. In darkness, our visual system is organized to give us a high sensitivity to light. This is partly achieved at the cost of other visual capacities. In darkness we (also) use a special and highly sensitive type of receptor cell (*rods*) and, when seeing only with the rods at levels of deep darkness, we have a low visual acuity and no colour vision.

In daylight, our vision of lightness contrast (1) and our vision of colours (2) are, of course, used simultaneously and in parallel. However, there are several reasons for considering these two daylight systems as separate functional entities:

perceptions of lightness contrast may become dissociated from those of colour after certain kinds of brain damage (Section 6.3.1);

to an important extent the two systems use different neuronal networks and, at least partly, their messages travel from eyes to brain along different portions of the visual pathway (Section 4.6.3);

only two of the three daylight sensors (L and M cones) are used by both systems.

Thanks to the parallel analysis of lightness contrast and colour, we may experience both a black-and-white photograph and a colour picture as being useful and adequate representations of the real world. Thanks to our colour vision, we may very rapidly discover a group of blue bellflowers in a meadow (Plate 4.1a), but with a little more time and patience we may also find them in a monochromatic version of the same photograph (Plate 4.1b). For a single bellflower, its characteristic shape is easily seen and recognized in a colour photograph (Plate 4.2a), in a grey-tone image (b), and in a drawing showing only the lines of contrast-borders, separating regions of different lightness (c).

(a)   (b)

Plate 4.1 Blue bellflowers (*Campanula*) on a meadow, the same photograph reproduced with (a) and without (b) colour. A black and white version of this figure will appear in some formats. For the colour version, please refer to the plate section.

(a)   (b)

Plate 4.2 The same photograph of a blue bellflower (*Campanula*) shown in full colour (a), in grey-tones (b), and after keeping only the (emphasized) borderlines between regions with a high lightness contrast (c). A black and white version of this figure will appear in some formats. For the colour version, please refer to the plate section.

(c)

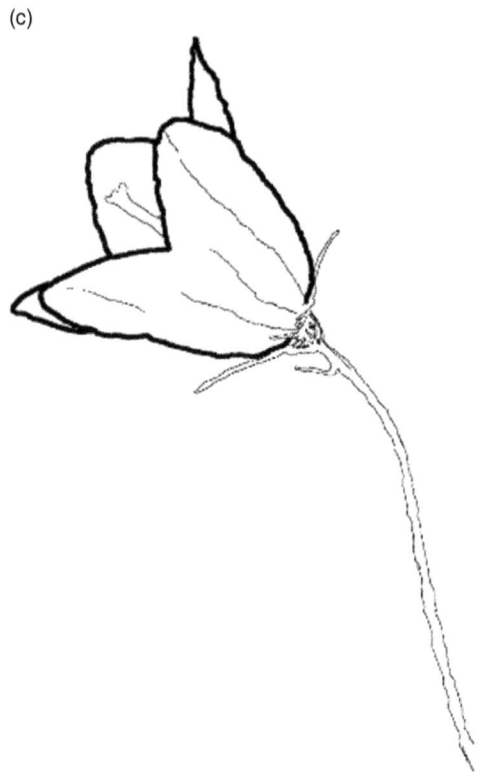

Plate 4.2 (*cont.*)

When trying to understand the relative roles of our 'luminosity' and 'colour' vision, it is particularly interesting to study visual functions in situations with objects of equal luminance, differing only in colour.[95] In the past, it was sometimes suggested that contrast edges due to 'pure' colour contrast would generally be invisible; modern experiments and observations have shown that this is not the case. In the absence of any luminance differences, natural objects may often become delineated purely due their differences in colour.[96] The luminosity and colour vision systems provide parallel functions with regard to the perception of movement and the analysis of shape and texture in objects. However, the two systems differ in threshold and in their capacities for detail detection. For the perception of contrast edges, pure chromatic stimuli require a higher contrast and give a lower spatial resolution than is the case for pure luminosity stimuli.[97]

How are our eyes and brains organized to be able to deal with our three visual systems?

4.2 Eye

4.2.1 Refraction and image projection

The human eye is a slightly flattened sphere with a diameter of about 2 cm (Figure 4.1). Its wall mainly consists of a tough layer of connective tissue (*sclera*), and the inside is covered by a layer of darkly pigmented cells which help to make the ocular wall highly impermeable to light. In front, the eye carries a lens system through which an image of the surroundings is projected into the eye, upside down and left–right mirrored. Most of the image refraction takes place at the transition between air and the curved and translucent front 'window' of the eye,

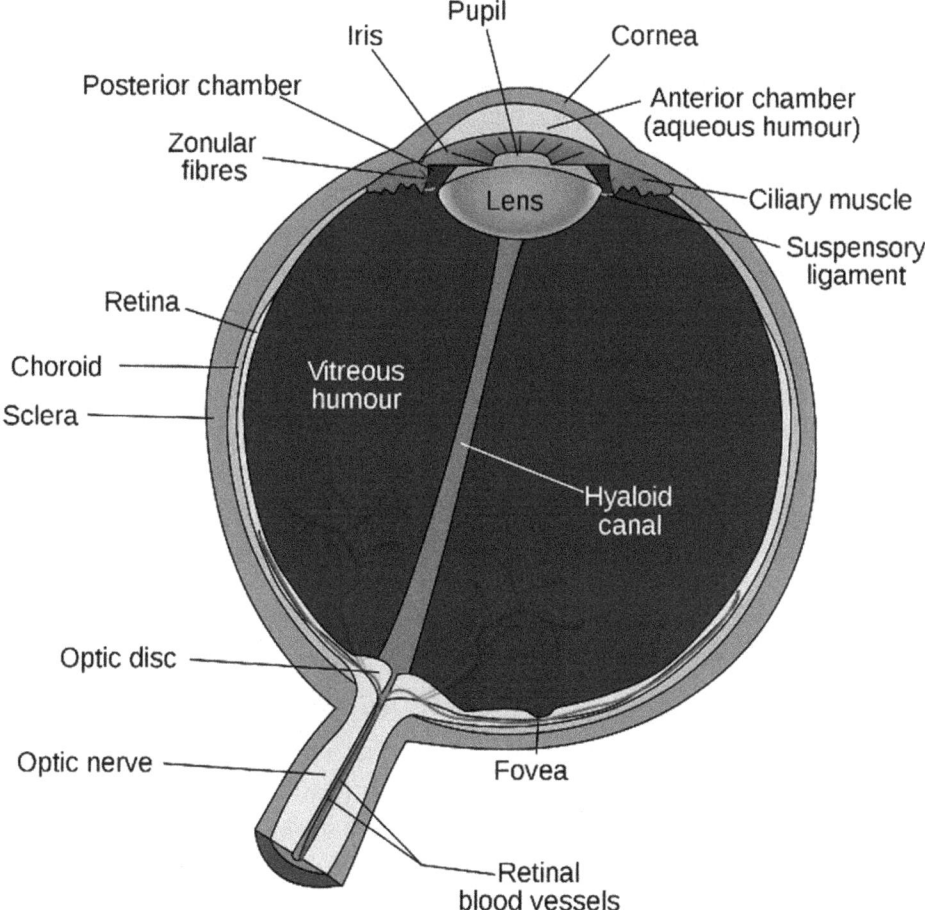

Figure 4.1 Anatomy of the human eye.
Source: http://en.wikipedia.org/wiki/File:Schematic_diagram_of_the_human_eye_en.svg.

its *cornea*. After passing through an anterior chamber filled with liquid, the light will be further refracted at the ocular *lens*. Behind the lens, the eyeball is filled with a kind of jelly (*vitreous body*), which is normally transparent for all the wavelengths of light. Ultimately, the image will be projected onto the light-sensitive tissue (*retina*) covering the inner back wall of the eye.

The ocular lens is elastic, and its shape may be changed for the focussing of images. At rest, the lens is normally set for observing distant things. When looking at items closer to the eye, a tiny intraocular muscle is activated which allows the lens to become more curved, so as to keep the retinal image in focus (*accommodation*). In older people, the lens becomes less elastic and, ultimately, it will not accommodate properly for the viewing of nearby objects; then the time has come for acquiring reading glasses.

4.2.2 Diaphragm of the eye: the iris

This is the only component inside the eye whose behaviour is directly visible to the outside. The iris is a more or less pigmented sheet (brown or blue eyes), with a circular shape and central hole (the *pupil*) for the entry of light into the eye. The size of the pupil is continuously regulated using two different intraocular muscles: a circular muscle making the pupil smaller, and a radial muscle making it larger. As one might expect, the brain controls these muscles such that the pupil generally becomes larger in darkness and smaller in bright daylight, i.e. pupil reactions help to adapt the eye to varying light intensities. In addition, the diameter of the pupil is important for the depth of focus of the eye, i.e. for the range of distances in front of the eyes at which objects are simultaneously in focus and seen as sharp. For a given focussing distance, the depth of focus becomes larger with a narrow pupil than with a wider one.

4.2.3 Sensor and processor: the retina

More than half of the inner back side of the eye is covered by the *retina* (Figure 4.2), an extremely complex tissue that contains the light-sensitive receptor cells, the *rods* and the *cones*. The total area of the retina is about 1094 mm^2 and it is about 80–320 μm thick. Besides the receptor cells, the retina contains large numbers of nerve cells (*neurones*) and supporting cells (*glia*). Embryologically, the retina is actually a part of the brain which has happened to land in the periphery. Some of the retinal neurones, the *ganglion cells*, have long nerve fibres which become bundled together into the optic nerve, transmitting signals from the eye to the brain for further use and processing (Figures 4.1, 4.2;

see Plate 4.4). As already discussed in relation to the 'blind spot' (Section 3.3.4), the retina lacks receptor cells at the optic disc, the site at which the optic nerve exits the eye (Figure 4.1).

4.2.4 The visual receptor cells

The retina contains three main kinds of light-sensitive cells, including two kinds of specialized visual receptor cells (Figures 4.2 and 4.3):

1. *Cones*, the receptor cells we use for both luminosity and colour vision in daylight, e.g. for seeing small details and colours. Each eye contains about 6 million cones.

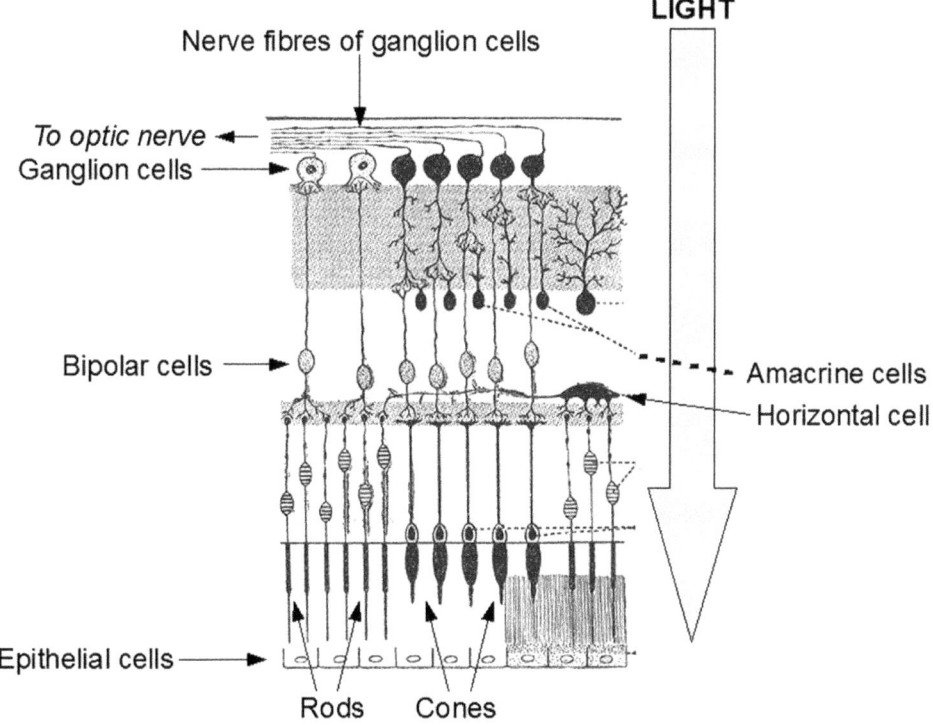

Figure 4.2 Structure of the retina, as seen in a vertical section through the tissue (thickness about 0.1–0.3 mm). The light enters from above, and the outer wall of the eye is below (epithelial cells). The diagram shows the two main types of receptor cell (rods and cones) and the four main kinds of nerve cell (bipolar, horizontal, amacrine, and ganglion cells). Only the ganglion cells have nerve fibres (axons) going to the brain. The other retinal cells exert all their visual functions via their direct and/or indirect influence on the activity of ganglion cells.
Source: http://en.wikipedia.org/wiki/File:Gray882.png.

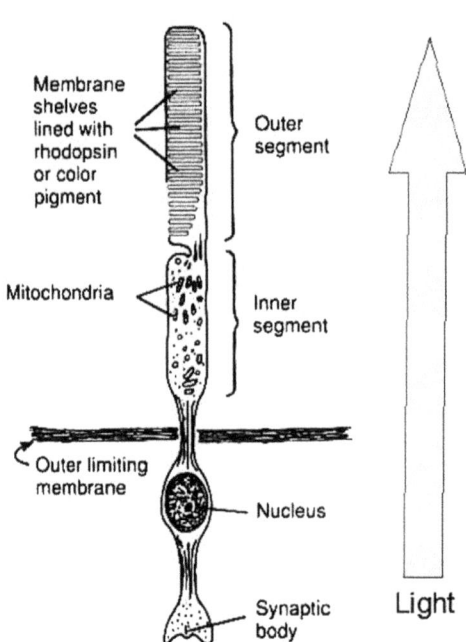

Membrane shelves lined with rhodopsin or color pigment

Outer segment

Mitochondria

Inner segment

Outer limiting membrane

Nucleus

Synaptic body

Light

Figure 4.3 Ultrastructure of visual receptor cells in vertebrate eyes (cones and rods). Within the same retinal region, cones are shorter than the rods. The light sensitive visual pigments are localized in the *outer segments* with their many transverse membranes. At their innermost ends, the receptor cells are synaptically connected to bipolar and horizontal nerve cells (cf. Figure 4.2).
Source: http://en.wikipedia.org/wiki/File:Rod%26Cone.jpg.

2. *Rods*, the receptor cells we use for sight in darkness. They are more light sensitive than the cones, and also much more numerous. Each eye contains about 120 million rods.

Rods and cones have a similar microscopic structure (Figure 4.3): they are long and thin with a diameter of only a few micrometres (1 μm = 1/1000 mm). Each cell has two main parts: an outer segment containing rows of transverse discs with light-sensitive molecules, and inner portions containing the nucleus and general cell components for energy production and protein synthesis. In addition, the innermost part has functional contacts (*synapses*) with retinal nerve cells. The rods are often about 50 μm long and, within a given region of the retina, they are typically a little longer than the cones; hence, the naming of the two kinds of cell. The light-sensitive portions of the visual receptor cells have an intriguing localization, close to the ocular wall, which means that the light has to pass through layers of nerve cells and inner portions of receptor cells before it reaches the light-sensitive molecules of the rods and cones.

Besides the visual receptor cells, the eye contains still another kind of light-sensitive element:

3. *Light-sensitive ganglion cells*, a strange kind of nerve cell that was discovered rather recently. These cells are particularly sensitive to the violet/blue light of short wavelengths and, via the optic nerve, they send their nerve fibres to the brain stem. There they have an influence on neural networks of importance for diurnal rhythms, i.e. on how bodily functions vary with the time of day and night. Thus, a main task of these cells seems to be to help synchronize our internal clock with the sun's daily cycle. Furthermore, the light-sensitive ganglion cells help to regulate the size of the pupil, making it smaller in strong light. Normally, these very specialized cells probably do not have any major role in the reception and processing of visual images. However, they might possibly play a role as a contingency device, enabling persons lacking rods and cones to perceive a difference between light and darkness. Nerve cells mentioned further on in this book do not belong to this exotic type.

4.3 Cones

4.3.1 The three kinds of cones

In daylight, the light processing of the eye depends on three different kinds of cones, each one having different wavelength characteristics. The properties of these receptor cells have been determined using psychophysical measurements in normal trichromats as well as in colour-blinds lacking one of the cone categories. In addition, measurements have been done at the level of single receptor cells. The three categories of cones have been named in accordance with their light sensitivities at different wavelengths (Plate 4.3; for further details, see Sharpe *et al.*, 1999):

- *L cones* have their highest sensitivity and light absorption at long wavelengths (maximum normally at about 560–570 nm);
- *M cones* have their highest sensitivity and light absorption at somewhat more intermediate (middle) wavelengths (maximum normally at about 530–545 nm);
- *S cones* have their highest sensitivity and light absorption at short wavelengths (maximum normally at about 420–440 nm).

Previously, L, M and S cones were often referred to as, respectively, the red, green and blue (-sensitive) cones. However, such a colour terminology is quite misleading because each category of cones is light sensitive over a much wider

(a)

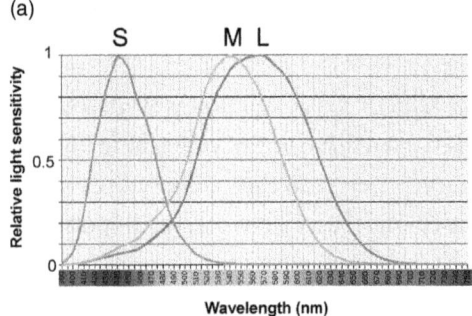

Wavelength (nm)

Plate 4.3 (a) The relative sensitivity to light of different wavelengths for the three kinds of cones in human eyes: S cones for short wavelengths, M cones for middle wavelengths, and L cones for long wavelengths. X-scale 390-750 nm. Data from Stockman & Sharpe (2000): "*cone fundamentals*", mainly derived from psychophysical measurements.
Source: http://commons.wikimedia.org/wiki/File:Spectrale_gevoeligheid_kegeltjes.png.
Image by Koenb.

(b)

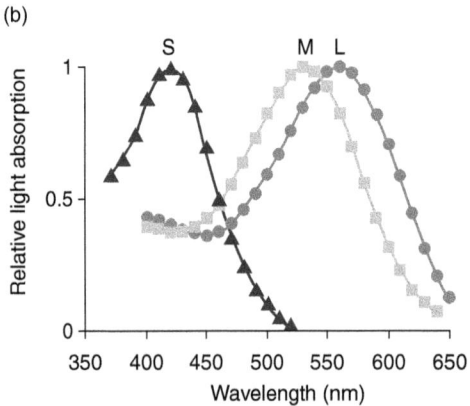

(b) Measurements of the relative light absorption at different wavelengths for the three types of human cones. Data from Dartnall et al. (1983). Note that, in both panels, data were normalized per type of cone; see text for details concerning quantitative differences between the respective cone populations.
Data from Dartnall *et al.* (1983), see http://www.cvrl.org/.
A black and white version of this figure will appear in some formats. For the colour version, please refer to the plate section.

region of the spectrum than a single hue (Plate 4.3a). In particular, the L and M cones have rather similar characteristics and they may each react to almost any wavelength across the spectrum, although with a different relative sensitivity for each wavelength (and, for the M cones, with a lack of sensitivity at the very longest 'red' wavelengths). Furthermore, the so-called 'red' cones (the L cones) actually do not have their maximum sensitivity in red portions of the spectrum but rather in its yellow-green region (Plate 4.3a).

Using microscopes and modern imaging techniques of *adaptive optics*, individual cones may even be seen and counted in the retina of living human eyes.[98] In such optical images, the three kinds of cone may be identified using methods of *selective light adaptation*, utilizing the cone differences in light absorption for different wavelengths (Plate 4.3b). When light is absorbed by the visual pigment of a cone, the pigment becomes *bleached*, i.e. it absorbs less light than before and looks paler (see Section 4.5.1 for the associated biochemistry). A bleaching light of 550 nm will affect M and L cones while leaving the S cones largely unchanged. A wavelength of 650 nm will mainly bleach the L cones, and with 470 nm the M cones are bleached more than the L cones. Using these methods for recognizing and counting the various types of cones, it has been found that their relative numbers are very different: the L cones are typically those most numerous (on average *c.* 65%), the M cones are often about half as common (*c.* 30%), and the S cones are remarkably scarce (*c.* 5%).[99] However, a considerable variation was found between individuals (for further comments, see Section 4.6.5).

In strong daylight, cones are the only receptor cells available for vision. As mentioned above (Section 4.1), these cells then have two parallel tasks: (1) to provide data to enable us to see and recognize small details in patterns of lightness contrast; (2) to provide data for our colour vision. Due to this parallel processing of visual images, we may combine colour vision with a high degree of resolution and sharpness.[100]

4.3.2 Cones and the sharp vision of lightness contrast

Lightness contrast is analysed using only the L and M cones; the S cones are not employed for such tasks, they are only used for colour vision. Furthermore, for the analysis of lightness differences, L and M cones are used together, as if being equivalent. Their distribution across the retina is clearly related to their major tasks in providing sharp vision. As we direct our eyes toward an interesting detail in our visual field, the image of that detail will be projected onto the specialized retinal region called the *fovea* (Figure 4.1). In this region, the L and M cones have their greatest concentration in terms of numbers/mm^2, which is advantageous for visual acuity. In central portions of the fovea, there are only cones and no rods. In the innermost foveal region, even the S cones are missing,[101] allowing the concentration of L and M cones to become even higher and acuity to become maximized.[102] Furthermore, in the fovea, the degree of optical resolution is additionally enhanced by other means: the sensory cells are not covered by the layers of retinal nerve cells only in this region. Here, the neuronal networks have been bent aside and the receptor cells may be reached by the light entering the eye in a more direct and undisturbed manner than elsewhere in the retina (Figure 4.2). The word 'fovea' means a pit or small cavity, and this refers to the small deepening at this site, due to the sideways displacement of neuronal structures.[103] From the fovea and out toward the periphery, the density of cones rapidly decreases, as does the daylight visual acuity (Figure 3.2). However, some cones are to be found almost anywhere in the retina, which is necessary for providing sufficient peripheral vision also in broad daylight.

It might seem strange that the nervous system does not use all the three categories of cones for sharp vision. However, from an optical point of view, this 'choice' is advantageous. In a lens system, the degree of refraction is greater for shorter than for longer wavelengths (Plate 2.1b), and lightness contrasts within an image would become increasingly blurred for an increasing range of differently refracted wavelengths.[104] Within our retina, this kind of blurring due to *chromatic aberration* is diminished by using mainly the longer wavelengths, i.e. those most effectively activating the L and M cones, for the analysis of lightness contrast. Furthermore, the potential blurring effects of the shorter wavelengths are diminished by limiting their access to the foveal region: this retinal portion is covered by a substance acting as a yellow optical filter. This yellow compound, *xanthophyll*, also absorbs much of the ultraviolet radiation, i.e. it acts as protective 'sunglasses' for the most important region of the retina. The xanthophyll-pigmented region is known as the 'yellow spot' or *macula lutea*, and it is about 3–4 mm in diameter, a little larger than the fovea itself.

4.3.3 Cones and colour vision

When looking at an isolated colour sample, our perception of colour depends on the *relative* degree of activation of the L, M and S cones within the corresponding retinal area (cf. Plate 4.3a). If the cones are activated by a light containing, like sunlight, similar amounts of all the visible wavelengths, then we see the sample as white, grey or black. This achromatic perception arises when each one of the three cone categories is activated to about the same degree. As one might see in Plate 4.3a, this state cannot be achieved with light on only one wavelength; hence, normally no achromatic hues are seen in a sunlight spectrum (Plate 2.4). However, a homogeneous activation of all the three types of cones may be achieved using only two well-chosen wavelengths of light, i.e. the light of two complementary colours. Looking at the wavelength characteristics of the cones, it becomes understandable that, for instance, yellow and blue are complementary colours: yellow would mainly activate the M and L cones and blue would mainly activate the S cones. Hence, with appropriate intensities, monochromatic yellow and blue light might together produce a similar degree of activation of all three cones, producing the perception of white/grey light.

Colours are generally seen when the S, M and L cones are activated to different degrees, and the identity of the perceived colour is then primarily dependent on the pattern of differential cone activation (cf. Plate 4.3a). Light with short wavelengths will mainly activate the S cones, which leads to perceptions of blue/violet. Light with many long but few short wavelengths might activate mainly M and L cones but few S cones: this might lead to a perception of yellow. Light dominated by the longest visible wavelengths and containing little else would activate L cones more than any of the others, and this might lead to the perception of red.

One intriguing detail concerns the relative sensitivities and influences of L versus M cones at the shortest visible wavelengths. Not only are colours seen as red for long wavelengths, but also within the violet shortwave end of the spectrum hues are perceived as being reddish. For longwave red colours, L cones are clearly more activated than M cones, which apparently leads to the perception of 'red'. What, then, is the cause for a reddish impression at the very shortest visible wavelengths? Are there also, within this spectral region, mechanisms that might enable the signals from L cones somehow to dominate over the signals from M cones, thereby generating the reddish tinge of violet?[105]

As mentioned above, the geometrical resolution for colour contrast is not very great. This is, for instance, illustrated by the fact that we never consciously notice that we are are 'blue-blind' within the most central portion of our visual field, the part where S cones are lacking. This blue-blind region has a diameter of about 0.35°, which would obscure about two-thirds of a full moon. In our normal daily vision, this is of no importance; the brain continuously fills in appropriate colours in a manner analogous to (and less drastic than) its online retouching of the blind spot (Plate 3.6). The central blue-blindness might lead to sensory consequences only in some very special situations (and it can, of course, be observed in specialized measurements). One example: the central blue-blindness might impair the results when searching for small items against a general blue background, e.g. from an aircraft flying over water.[106] Another curious example was reported during World War I.[107] In the beginning, the

French soldiers had red caps, and they were often shot as soon as their heads emerged above the trenches. The colour of the caps were then changed to blue, and thereafter many fewer soldiers were shot under similar circumstances. The tentative explanation was that, due to the central blue-blindness, it might be very difficult to fixate a tiny blue object at a large distance. In order to see the minute blue target as clearly as possible, a sniper would probably have had to direct his eyes a little to the side of the blue cap itself, and this might have disturbed him when aiming his gun.

4.3.4 Cone properties and the wavelength/intensity distributions in natural scenes

Our visual capacities have, of course, developed in response to properties of the surrounding physical world. Hence, it is of interest to consider in which ways our sensory capacities have become optimized for the gathering of information from our natural surroundings. Such comparisons between our visual system and physical properties of natural scenes have been statistically explored in various recent studies.[108] As already pointed out (Section 2.4, Figure 2.2), the wavelength range used for our sense of vision represents one initial optimization: the visible wavelengths coincide with those for which sunlight has its greatest energy content when reaching the earth. Furthermore, an analysis of isoluminance chromatic contrast in natural scenes revealed that, independently of each other, visible chromatic and luminance edges are both important, i.e. the human kind of colour vision gives us increased possibilities for perceiving the outlines of surrounding objects.[109] For the discovery of such chromatic contrast, signals from different kinds of cones are processed such that relevant differences between them are clearly distinguished (see below, Section 4.6.5 and note 117).

4.4 Rods

4.4.1 Seeing in darkness

In very dark surroundings, when seeing only with rods, our vision differs from that of normal daylight in several important ways, including:

- In darkness we cannot see small details, our visual acuity is considerably lower.
- In darkness our light sensitivity is relatively low at the centre of our visual field, i.e. at sites corresponding to our fovea we are 'nightblind'. Our highest

sensitivity to light occurs if looking a little 'to the side', a known trick for, for instance, astronomers gazing at weak stars.

– In darkness we see no colours in those regions of our visual field where light is subthreshold for the cones. However, within such dark areas some details might radiate sufficient amounts of light to activate cones and their colour mechanisms. Thus, for instance, walking around outside at night, the landscape might be monochromatically grey/black while the lighted windows of houses look brightly yellow.

The sharpness of our night vision is low in spite of the enormous numbers of available rods. At night, light sensitivity is prioritized above sharpness, and this is achieved by connecting many rods to each nerve cell. Our relative 'nightblindness' at the very centre of the visual field is simply due to the lack of rods in this part of the retina (rod-free central portion of the fovea). The relative density of rods is maximal in a circle at about 18° around the centre (the fovea), and it declines very rapidly toward the centre and more gradually toward the periphery. Our complete colour blindness on dark nights, when having only rods available for our vision, is a consequence of having only one type of rod. In order to perceive differences between lights of different wavelength composition, one must use at least two types of receptor cells with differing wavelength sensitivities.

4.4.2 Sensitivity of rods to different wavelengths of light

Using brief flashes, the sensitivity of the eye may be measured for light of different wavelengths. The relationship between wavelength and light sensitivity is somewhat different depending on whether the eye is adapted for use in daylight or darkness.

As is demonstrated in Figure 4.4, the dark-adapted eye is less sensitive to longwave 'red' light and the maximal sensitivity occurs at a shorter wavelength in darkness (c. 507 nm) than in bright daylight (c. 555 nm). These differences reflect the different wavelength characteristics of rods versus cones. The relatively great sensitivity of rods for short wavelengths gives rise to a change in the relative brightness of blue versus green colours: against the background of green grass, blue flowers look lighter at night than in daytime. This adaptation-related change in the perceived luminosity of different colours is called the 'Purkinje shift', after the Czech physiologist Johannes Evangelista von Purkinje (1787–1869).

Direct measurements on single rods have confirmed that their wavelength sensitivity corresponds to the *scotopic* sensitivity characteristics for the eye

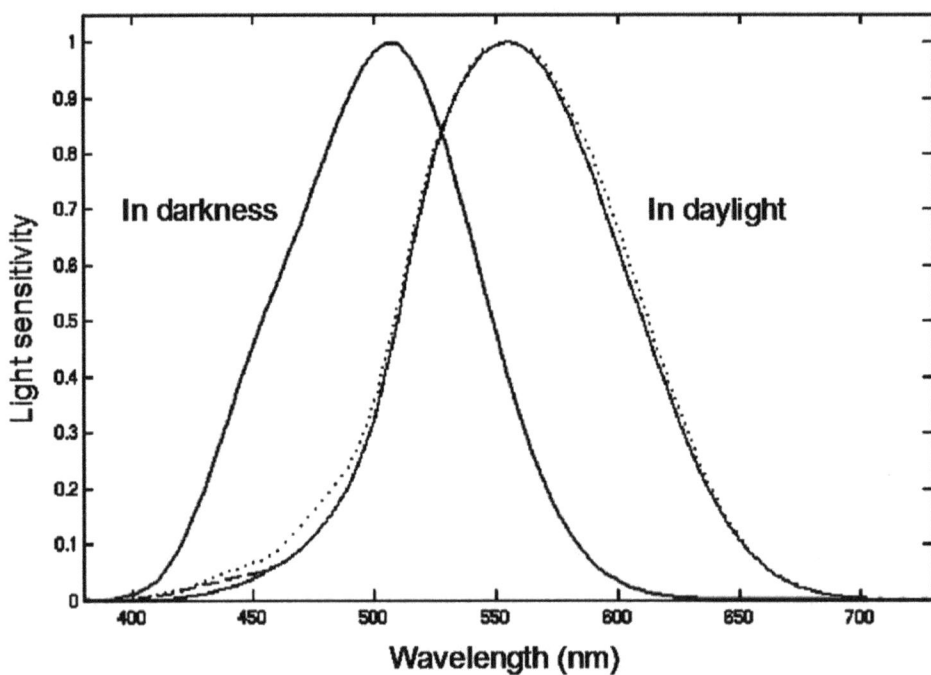

Figure 4.4 Sensitivity of the eye to light of different wavelengths in full daylight (*photopic vision*) and in darkness (*scotopic vision*). The scotopic curve ("*In darkness*") corresponds to the light sensitivity of the rods.
Source: http://en.wikipedia.org/wiki/File:Luminosity.png.

(Figure 4.4, left curve). Furthermore, it has been shown that this sensitivity profile largely corresponds to the absorption properties for the light-sensitive molecules of the rods, their *visual pigment*. The wavelengths that are most readily absorbed are also those with the greatest influence on the signalling of the receptor cells. This is true for both rods and cones (cf. Plate 4.3), and the relation between absorption and sensitivity is of practical use when comparing eyes of different animal species: at the level of single cells, it is faster and technically less demanding to measure light absorption than to record the cellular signalling in response to different qualities of light.[110]

Under conditions of maximal dark adaptation and for light of optimal wavelengths, a single rod might become activated by a single photon, i.e. by the smallest possible quantity of light. Furthermore, rods may add (integrate) the effects of successive photons during a relatively long period of time, up to about 100 ms. This does not mean that rods will usually react to every single photon that hits the eye; for instance, very many of the photons entering the eye (perhaps 80%) will never be 'seen' due to their absorption by other intervening tissues before reaching the visual receptors.

4.4.3 The possible role of rods in colour vision

Over a certain range of lower light intensities, both rods and cones are capable of reacting to changes in luminosity; the light is then strong enough for the cones and not too strong for the rods (Section C.5). Our visual system then works under so-called *mesopic* conditions, i.e. a stage intermediate between the *scotopic* vision in darkness and the *photopic* daytime vision. In mesopic vision we have a simultaneous access to four different kinds of light receptor cells. How does this affect our colour vision?

The visual pigment of the rods (*rhodopsin*) has its greatest sensitivities for wavelengths between the optimal ones for S and M cones (cf. Figure 4.4 'In darkness' versus Plate 4.3). Theoretically, one might therefore possibly expect to find a tetrachromatic colour vision under mesopic conditions, i.e. a more advanced variety than the trichromatic vision of normal daylight. For people with normal trichromatic daytime colour vision, mesopic conditions are indeed associated with complex changes in colour vision as well as in luminosity sensitivities, but the effects are not sufficiently drastic and clearcut to be definable as a change toward tetrachromatic vision. However, rods might play a role for enhancing the colour discrimination of (some) colour-blind subjects under mesopic conditions (Section 5.5.2).

4.5 Chemical mechanisms of light sensitivity

4.5.1 The visual pigments

In visual receptor cells, the rods and cones, the outer segments contain large amounts of *visual pigment*, a substance whose properties may change when influenced by light. The visual pigment is inserted into rows of membrane lamellas in the outer receptor segments (Figure 4.3). Human eyes have four different visual pigments, one for the rods and one for each of the three kinds of cone. A visual pigment consists of two interconnected molecules: *opsin* and *retinal*. Retinal is a rather small molecule which is fabricated from vitamin A, and opsins are proteins and much larger. The opsins determine the wavelength sensitivity (cf. Plate 4.3) of the associated *chromophore* of the pigment, the retinal. The four kinds of human visual pigment all use the same kind of retinal, and each one has its own characteristic type of opsin.

Like all proteins, opsins consist of long chains of a standardized kind of molecular 'building block', the amino acids. There are, for this purpose, 20 different kinds of amino acid, and the functional properties of a protein depend on the precise sequence and identity of its amino acids. For each protein of our body, this sequence is encoded and stored in the DNA molecules of our genes. Each gene encodes at least one protein, and there is a separate gene for each one of the four kinds of opsin. Our various opsins are all clearly similar to each other in their amino acid composition and sequence. Such similarities are particularly striking for the two kinds of longwave cones: the opsins of the L and M cones differ from each other in only 4% of all their 364 amino acids.

The other molecular part of the visual pigment, the retinal, consists of carbon, hydrogen and oxygen atoms. While keeping the same sequence within each retinal molecule, the constituent atoms might be linked together in geometrically different ways, producing retinals of different spatial shapes (*isomerism*). The shapes of molecules are extremely important for their functional properties. This is strikingly illustrated by the reactions of retinal when light is absorbed by a visual pigment: the shape of the associated retinal is then changed from a steric variety called *11-cis* to another one called *all-trans*. This altered shape of the retinal leads to a whole series of chemical processes which, ultimately, causes a change in the signalling from the receptor cell to its connected nerve cells (see next section for further details). Interestingly, these biochemical changes are also associated with rapid alterations in the light absorption characteristics of the visual pigment: its wavelength for maximum absorption is shifted and the pigment loses its original colour, a phenomenon called 'bleaching'. These changes may be directly observed with a microscope.

Visual receptor cells are all connected to extensions from one or several nerve cells (Figure 4.2). At such *synaptic* connecting sites, the membranes of the two cells are not directly fused but they are lying very close to each other. In such a *synapse*, the receptor cell may release a neuronal transmitter substance which affects the membrane and, thereby, the activity of the connected neurone. Rods and cones continuously release transmitter substance in darkness, which leads to a continuous activating or inhibiting influence on the various kinds of directly connected neurones (review: Wässle 2004). The primary effect of light is to *decrease* this release of transmitter from the receptor cells, thereby causing a *decreased* activation (or *decreased* inhibition) of the directly connected nerve cells. One of the possible advantages of such a paradoxical kind of reaction is that, as the receptor system is already up and running in dim light and darkness, it is continuously prepared for immediate reactions to small changes in light intensity, without any problems with thresholds or onset delays. For vision, it is essential that a change in light intensity produces a change in nerve cell activity, but the direction of the primary changes is less important. Due to interconnections between excitatory and inhibitory cells within vast neuronal networks, the activation of visual receptor cells might lead to an increased activity of some and a decreased activity of other neurones within the retina and the brain.

4.5.2 Dark and light adaptation

We may use our eyes in situations that vary enormously in the intensity of illumination. Outside, the light is about 1000 million times more intense on a sunny day than in a cloudy night with no moon.[111] Passing from bright daylight into a dark room, one might briefly become almost blind. Then the light sensitivity will gradually recover (*dark adaptation*), and after a few tens of seconds to minutes vision might return to normal. Similarly, going into broad outside daylight from the shadows of a dark room might also temporarily overload the eyes such that they become useless (often referred to as *bleaching adaptation*, i.e. changes associated with a bleaching of visual pigment). However, in this latter situation the recovery (*light adaptation* or *background adaptation*) is quite rapid, clearly faster than that for the dark adaptation.[112] To some extent, the visual adaptations to changes in light intensity depend on changes in the ocular diaphragm, the size of the pupil as regulated by the iris muscles. This mechanism works rather rapidly, but it can only change the quantities of light entering the eye by a factor of about 1:16 or less.[113] By far the most light and dark adaptation takes place due to chemical changes, altering the light sensitivity of the visual receptor cells.

When measuring the time course of dark adaptation, the participant first spends enough time in bright daylight to make the eyes appropriately light adapted. He/she is then placed in a dark room, and at different moments his/her light sensitivity is tested using flashes of varying intensity, to determine how strong the flashes have to be to become barely visible (*threshold intensity*). Plotting threshold intensity versus time for such an experiment will typically produce a biphasic curve like that of Figure 4.5. At first, there is a rapid fall of threshold (i.e. increase of sensitivity) during about 5–8 min. Then follows a second, more prolonged phase of threshold decrease which might last up to 40–50 min. The initial rapid phase of dark adaptation mainly reflects the adaptation of the cones; at these light intensities the rods are mostly over-stimulated and rather useless. The slower second phase of the adaptation depends mainly on the rods; these intensities are mostly below the thresholds for the cones. The total duration of dark adaptation might seem impractically slow, but it fits rather well to the time it takes for the sun to set in the tropics (about 20–25 min close to the equator).

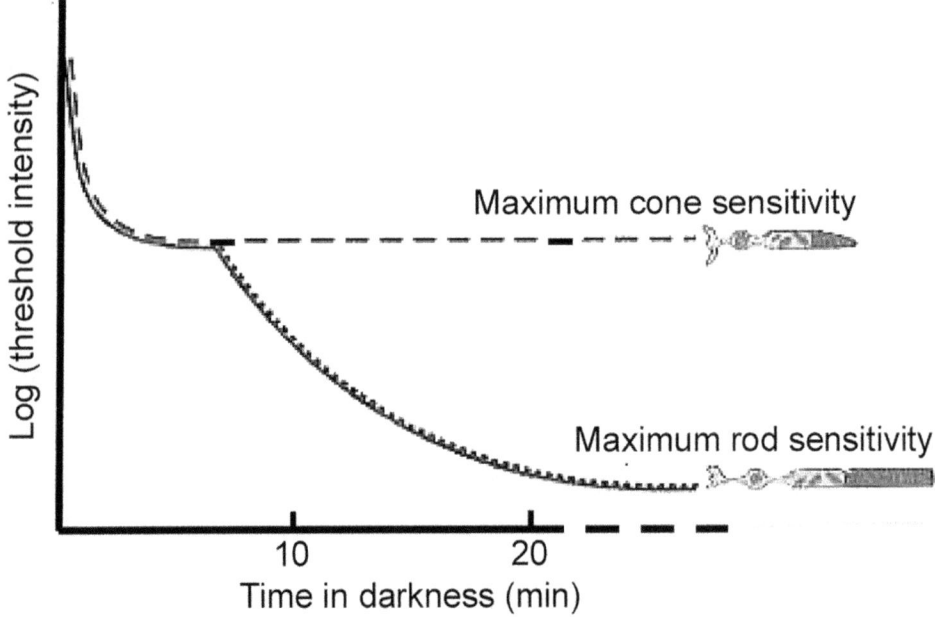

Figure 4.5 Approximate time course of human dark adaptation (schematic representation). The participant at first spends sufficient time in normal daylight to become fully light adapted. The participant is then transferred to a dark room and, at different times thereafter, measurements are made with regard to the visual threshold, i.e. it is tested how strong a brief light has to be in order to become just barely visible. The y-scale is logarithmic. The threshold intensity may, for instance, decrease by a factor of about 100 during the initial cone phase and by about 1000 in the succeeding slower phase of rod adaptation.
Source: http://openwetware.org/wiki/Image:DarkAdaptation.jpg.

Within individual cones and rods, the absorption of light in the visual pigment will cause its retinal to change its shape from 11-cis to all-trans, and this causes a bleaching of the pigment and a series of changes in its biochemical properties. Via a rapid sequence of intervening steps, lasting only about 1 ms, the pigment reaches a stage in which it will trip off the cascade of biochemical events leading to changes in neuronal activation (see Section 4.5.1). This transient activating version of the pigment (in rods called metarhodopsin II) is soon, in ~1–2 s or less, further transformed into a non-activating form (metarhodopsin III) which will ultimately, after several minutes, be split up into its two main portions: opsin and all-trans-retinal. Strangely enough, essential steps in the subsequent recovery of light-sensitive visual pigment (*pigment regeneration*), takes place outside the receptor cells themselves. This concerns the transformation of light-insensitive all-trans-retinal to the light-sensitive 11-cis form, occurring in the so-called *epithelial cells* which cover the inside of the eye's outer wall, just outside the outer segments of the receptor cells

(Figure 4.2). Thus, the regeneration of retinal from its all-trans to its 11-cis shape requires that the retinal molecules are moved back and forth between two different classes of cells, the receptor and the epithelial cells. Such molecular movements take time, and this helps to explain why the dark adaptation is such a slow process. The regeneration of light-sensitive visual pigment is much faster for cones (~10 min) than for rods (at least 20–30 min).

The light sensitivity of a visual receptor cell varies with the concentration of its light-sensitive visual pigment. This concentration will automatically be set at different levels, depending on the general light intensity. For weak illumination, the turnover of retinal is slow enough to keep much of the visual pigment in its active form, which is beneficial for light sensitivity. The stronger the general light intensity, the more of the pigment molecules will become bleached and insensitive, and the receptor cells will become light adapted. The latter process is relatively fast, typically completed within a couple of minutes.

In addition to the importance of 11-cis visual pigment concentration for light sensitivity, other processes and mechanisms also play a role for the light/dark adaptation. After bleaching of visual pigment, the decreased light sensitivity of a receptor cell is partly due to direct 'desensitizing' influences of decomposition products of the photoactivated pigment (e.g. in rods a rather prolonged presence of metarhodopsin III, and of free opsin molecules which are not yet provided with light-sensitive 11-cis-retinal).[114] Furthermore, pupil changes play a role (see above), and there are also adaptation-associated changes in the organization and functions of the eye's neuronal networks. For instance, dark-adaption causes some of the nerve cells to receive synaptic inputs from larger groups of receptor cells, thus promoting a high light sensitivity at neuronal levels.

The mechanisms of dark/light adaptation are localized within each eye, and this makes it possible to let these processes take place in only one of the two eyes, e.g. using only one eye in bright daylight while keeping the other one covered in a dark-adapted state. The rods are relatively insensitive to longwave red light, and this makes it possible to use dark red glasses for keeping the rods in a dark-adapted state while spending some time in a brightly lit room. Such tricks were apparently used by British pilots during World War II, and by medical staff in old-fashioned X-ray examinations employing weakly fluorescent monitor screens.

4.5.3 Negative and positive afterimages

The phenomenon of negative afterimages (Section 3.1.2) depends on localized patterns of light adaptation among the cones. Staring for 20–30 s at a well-lit coloured picture makes different sets of cones light adapted for each one of the different coloured details in the picture. These adapted cones become less sensitive than the other ones (Section 4.5.2). As the eyes are then closed or directed toward a neutral scene (e.g. grey or white paper), the light-adapted cones will, for a limited period of time, have a different resting activity than that of other cones in the same region, as if the adapted cones now received less light. Therefore, the non-adapted cones will now dominate in the colour analysis, temporarily causing effects as if they received more light than the adapted cells. Hence, the temporary appearance of complementary colours in the negative afterimage. Precisely which colours will appear due to such an 'imbalance' between the activities of different sets of cones depends on how the colour information is processed by neuronal networks of retina and brain (Section 4.6.5). It is important, in this context, to remember that the cones are continuously releasing transmitter substances even in darkness.

The short-lasting positive afterimages (Section 3.1.2) give an impression of how long the neuronal activity, as evoked by a visual image, continues after the end of the external stimulus, i.e. these kinds of afterimages give information about important aspects of the dynamics of the visual system. In this context, it is important to consider the speed with which the receptor cells react to changes in light intensity. One might, for instance, measure the *flicker fusion frequency*, i.e. the maximal frequency at which a variation in light intensity still gives a variation in reactions of the visual system. Under similar and optimal conditions, this frequency is substantially higher for cones than for rods (e.g. ~80–90 Hz versus ~10–15 Hz).

4.6 From eye to brain

4.6.1 Structure and function of nerve cells

Nerve cells (*neurones*) come in many different shapes, but mostly one might recognize three main components (Figure 4.2):

1. The cell body (*soma*), often more or less spherical in shape, contains the nucleus with its genetic code and often receives synapses from other cells onto its membrane
2. One or (usually) several *dendrites*, fibre-shaped extensions, typically emanating from the cell body and provided with many side branches, receive synaptic connections from other nerve cells.
3. The *axon*, a fibre-shaped extension that may be very long, usually starts at the cell body, conducts signals (e.g. *nerve impulses*) from its parent neurone to other cells and is typically widely branched when it reaches its destination, there connecting to many other cells.

Neurones are used for the transmission and processing of information. This occurs in neuronal networks in which the activity of each neurone is influenced by that of many others. Each neurone is, as it were, a kind of processing centre in which hundreds or thousands of various incoming signals are gathered together. The incoming signals may be positive or negative or modulating, and the task of each neurone is to produce an outgoing signal which somehow reflects a balanced sum of all the incoming messages. This outgoing signal is then sent on to hundreds or thousands of other neurones, processed in a similar way in these cells, producing new outgoing signals, and so on. The properties of such a processing system depend very much on how its various components are interconnected, and these inter-neuronal links are, to a great extent, very precisely specified early on during the embryological development. Further-more, an essential property of our own neuronal networks is that their process-ing properties might be made to change due to various factors, including influences of the information stream itself; within strict (and still partly unknown) limits, neuronal systems are plastic and may learn from experience.

As already mentioned (Section 4.5.1), the signalling between receptor cells and nerve cells occurs via synaptic connections, and this is also true for the communi-cation between neurones. Almost all of these synapses are of a chemical type, i.e. the signalling cell releases a transmitter substance into the synapse, and this rapidly causes a change in the activation state and/or properties of the receiving cell. In most (but not all) kinds of neurones, a sufficiently strong and long-lasting synaptic activation will cause the receiving cell to send out a series of nerve impulses along its axon (see next paragraph), and these impulses will then arrive at the various synapses made by this axon on other neurones, influence these neurones via a release of synaptic transmitter, and so on. Most synapses belong to one of two main types: either their transmitter substance might cause an increased activation of the next cell (*excitatory synapse*), or it might make activation more difficult (*inhibitory synapse*). These two kinds of interaction are both about equally common. In addition, there are *modulatory synapses* of various kinds, in which the main effect is to change the processing properties of the receiving cell.

Important aspects of the activation state of neurones are directly associated with electrical phenomena, which can be recorded and measured in various ways, both for single cells and for cell populations. Using thin recording electrodes and suitable electrical amplifiers one might, for instance, observe the *nerve impulses* that neurones produce when sufficiently excited. Each single nerve impulse (*action potential*) shows up as a rather large shift of voltage across the cell membrane (~0.1 V) and it lasts for only about 1 ms (= 1/1000 second). Typically, neuronal action potentials start off at the beginning of the axon, close to the cell body, and they then travel along the whole axon and cause transmitter to be released at synapses of the terminal axonal branches. At rest, a neurone might produce very few or no nerve impulses per second. When strongly activated, nerve impulses might occur at rates of tens to hundreds per second. For nerve impulses and associated bioelectric phenomena, the currents are carried by ions swimming around in water, which means that action potentials move at exceedingly slow speeds as compared to those of electrical pulses along metal conductors. The velocity of action potentials travelling along a nerve fibre (an axon) is of the order of 1 to <100 m/s, whereas the speed of electrical pulses moving along metal wires may be c 200 000 000 m/s.

4.6.2 Nerve cells of the retina

The visual pathway begins with the optic nerves from the eyes, but these nerves do not start off directly at the receptor cells (i.e. the rods and cones). The visual information travelling along the nerve fibres from the eye to the brain has already been thoroughly processed by retinal nerve cells. As is schematically shown in Figure 4.2, there are four major types of neurones in the retina. Two of these categories concern cells with a 'vertical' orientation, from the receptor cells toward the optic nerve:

- *bipolar cells*, coupled between the receptor cells and the *ganglion cells* (directly or via *amacrine cells*);
- *ganglion cells*, whose axons enter the optic nerve, continuing to the brain.

Two other cell types are organized 'horizontally', i.e. they interconnect different parts of the retina, serving important aspects of the intra-retinal visual processing. These cells include:

- *horizontal cells*, at the level of the synaptic contacts between receptor cells and bipolar neurones, both of which also have synaptic contacts with horizontal cells.
- *amacrine cells*, at the level of the synaptic contacts between bipolar neurones and ganglion cells, both of which also have synaptic contacts with amacrine cells.

Each one of these four major kinds of intra-retinal neurones have been categorized into several subtypes, e.g. related to differences in size, in intra-retinal position, in the number of contacts with other cells, in which types of neurones or receptor cells are contacted, etc. Thus, the retina is considered to contain about 20 different types of ganglion cells, 9 kinds of bipolars, 2–4 kinds of horizontal cells and about 22 kinds of amacrine cells which illustrates its great complexity.

4.6.3 The visual pathway

For each eye, the retina contains several millions of receptor cells and neurones. However, the optic nerve contains only about 1.2 million nerve fibres, each one being an axon from a retinal ganglion cell. All these fibres, bundled into the optic nerve, exit the eye at the *papilla* (*optic disc*, Figure 4.1). On their way toward the brain, the two optic nerves coalesce and seem to cross at the so-called *optic chiasma* (Plate 4.4); after the crossing the two nerves are referred to as *optic tracts*. At the chiasma there is no fusion of individual nerve fibres, but the fibres are redistributed such that, for each eye, all the axons from the left half of the retina proceed toward the left half of the brain, and vice versa. When the outside world is optically projected onto the retina, it becomes mirrored and turned upside down. Hence, the left hemisphere and the left half-eye will be concerned with the right-side half of the visual field, and vice versa. The left hemisphere is also the one most directly in command of muscles on the right side of the body; thus, it seems very appropriate that it also receives, from both eyes, the visual signals from the same half of the outside world.

Axons of the optic nerve/tract themselves never reach the cortical regions of the hemispheres. Some of the axons are connected to structures in the brain stem, serving visuomotor functions and internal functions of the

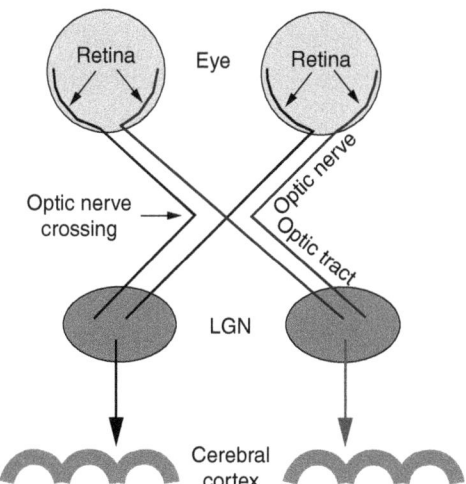

Plate 4.4 Structural features of the visual pathway and of the "crossing" of the optic nerve (*optic chiasma*). In the "interbrain" (*diencephalon*, at upper end of brain stem), nerve fibres of the optic nerve make synaptic contacts with nerve cells of a group called *Lateral Geniculate Nucleus* (*LGN*). Nerve fibres from the LGN transmit the visual information to the cerebral cortex. For both eyes, visual signals from the left side of the retina are transmitted to the left cerebral hemisphere, and vice versa. A black and white version of this figure will appear in some formats. For the colour version, please refer to the plate section.

eye, such as focussing and regulating the pupil size. However, most of the ocular axons are landing on neurones in the *lateral geniculate nucleus* (*LGN*), a collection of nerve cells in the *diencephalon*.[115] On each side, this intermediate station contains about as many cells as the number of nerve fibres from each eye (c. 1 million). The visual information is synaptically transmitted to the LGN neurones, most of whose axons proceed toward cortical regions at the back of the hemispheres, the so-called *primary visual cortex* (= *area V1* = *area 17* = *striate cortex*), containing several millions of nerve cells on each side. A small number of LGN axons proceed directly to other cortical areas.

Within the primary visual cortex, information from each position within the visual field is mapped out to a separate site in the cortex, and this is done with a joint cortical localization for corresponding positions from both eyes. This is possible thanks to the redistribution of the ocular axons in the chiasma, and it enables the cortical mechanisms to compare corresponding details in the images from both eyes. Such comparisons are essential for our binocular three-dimensional vision and binocular colour mixture (Section 4.6.5).

Studies on monkeys have shown that within the pathway from retina via LGN to V1 there is a certain degree of specialization related to the size of the nerve fibres. Thus a distinction is made between:

1. A *parvocellular* ('small-cellular') pathway with relatively thin retinal nerve fibres which connect to parvocellular layers of neurones in the LGN. These axons typically carry colour-specific information, i.e. reacting differently to the activities of different kinds of cones.
2. A *magnocellular* pathway, in which the retinal nerve fibres are relatively thick and fast-conducting. Retinal axons of this system connect to magno-cellular layers of neurones in LGN. These axons usually carry luminance-related information from different types of cones used together.

In comparison to the parvocellular system, cells of the magnocellular portion are typically colour blind, their receptive fields are large (i.e. low spatial resolution), and they have a high contrast sensitivity and rapid reactions. Differences between the parvo- and magnocellular portions of the visual pathway are partly preserved also beyond the LGN. Their respective LGN cells connect to different layers of the visual primary cortex, and their information is partly channelled to separate preferential locations also within still 'higher' regions of the visual system. The fine details and possible variability and overlap in these functional and anatomical subdivisions are still a matter for further experimental analysis. However, these findings illustrate the recurring phenomenon that, within the brain, complex functions may be split up into subfunctions which are dealt with in parallel in different places.

4.6.4 Neuronal receptive fields and the processing of lightness contrast

Visual neurones generally have a definable *receptive field*, i.e. a limited portion of the visual field within which an appropriate change of light intensity gives an alteration of neuronal impulse activity, for example, as explored using small spots of light projected in front of the eyes. Cells of the visual pathways are often spontaneously active and the effects of a spot of light might be excitatory or inhibitory. An excitatory effect is visible as an increased number of nerve impulses per second (*ON response*), and for inhibition the opposite behaviour is seen (*OFF response*; Figure 4.6).

For ganglion cells of the retina (i.e. for fibres of the optic nerve), the properties of the receptive fields are well known. These cells are relatively easily investigated, and their activities are, of course, of great importance as they provide the brain with all its primary visual information. The impulse activity of the ganglion cells does not change very much for a general alteration of light intensity across the

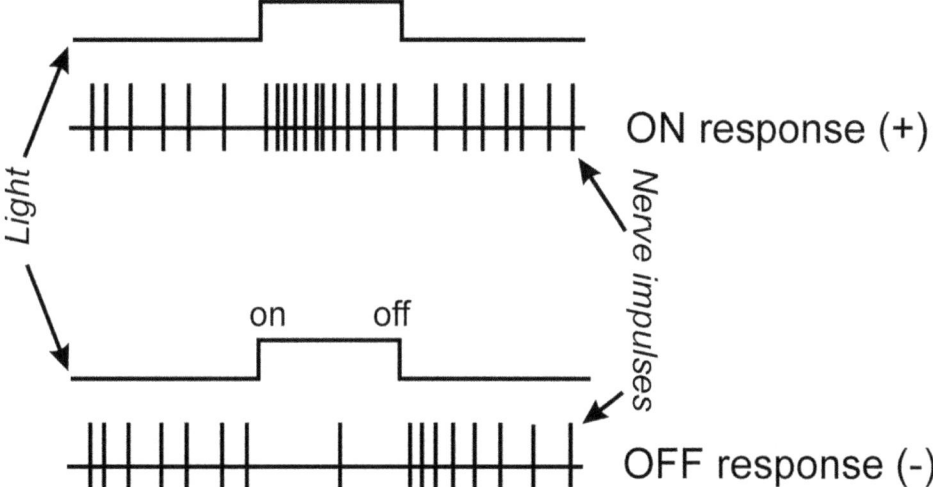

Figure 4.6 Scheme illustrating the kind of electrical activity that might be recorded from single neurones in eye or brain in response to spots or patterns of light. The vertical lines represent nerve impulses, short-lasting phenomena that might emerge in an active nerve cell, at higher rates the higher the level of cellular activation. Depending on how the nerve cell in question is coupled to the neuronal network, the visual stimulation might cause the cell to emit greater numbers of nerve impulses per unit time (stimulating effect, *ON response*), or the rate of nerve impulses might become reduced (inhibiting effect, *OFF response*).

whole visual field; our eyes and visual mechanisms are not used for assessing absolute and general light intensities. The main task of our visual system is to discover and recognize detailed patterns of *differences* in light intensity within the visual field, i.e. detecting regions of *lightness contrast*. This is how we can identify and recognize various objects in front of our eyes. The first steps in this luminosity contrast analysis occur already within the retina. How this happens becomes evident when mapping out the receptive fields of the ganglion cells (Figure 4.7). In such experiments, the activity of a single ganglion cell is recorded while the cell is influenced by a small spot of light at different sites within the visual field. Typically, this test light will cause an activation of the cell at some sites and an inhibition at other places. If the spatial distribution of these various response sites is mapped out, the receptive field turns out to be circular in shape with an orderly and concentric separation between the ON and the OFF sites. For about half of the cells, there is a central accumulation of ON sites and a periphery filled with OFF sites (*ON-centre cells*, Figure 4.7a); for other cells precisely the opposite is seen (*OFF-centre cells*, Figure 4.7b). Such concentric receptive fields provide an effective mechanism for detecting lightness contrast: the greatest change of neuronal activity would be produced by illuminating as much as possible of the central part of the field while leaving as much as possible of its peripheral parts in the dark (or vice versa; Figure 4.7c, d). Shining a test light across the whole receptive field would provoke only a weak response because the ON and the OFF reactions would largely cancel each other out. In practice, the weak responses to such general light stimuli usually take place in a direction set by the receptive field centre; e.g. a weak general ON response for an ON-centre cell.

The brain knows nothing about the activities of individual receptor cells in the retina; everything the visual brain knows has been reported to it by the ganglion cells, and their continuously transmitted messages contain the results of highly processed summaries of how the light affected different retinal populations of rods and cones. This complex processing has been done by networks of all the different types of retinal neurones together: bipolars, horizontals, amacrines and ganglion cells.

The sizes of the receptive fields are very different for different ganglion cells, and within the retina they are clearly related to local differences in visual acuity. In the central portions of the fovea, cones may even have a 1:1 relationship with their ganglion cells, i.e. the central portion of the respective receptive field might have a diameter corresponding to that of the outer segment of a single cone. In the most peripheral regions of the retina the receptive fields might instead be gigantic with huge numbers of receptor cells connected to single retinal ganglion cells. Between these extremes, numerous intermediate sizes of receptive fields occur.

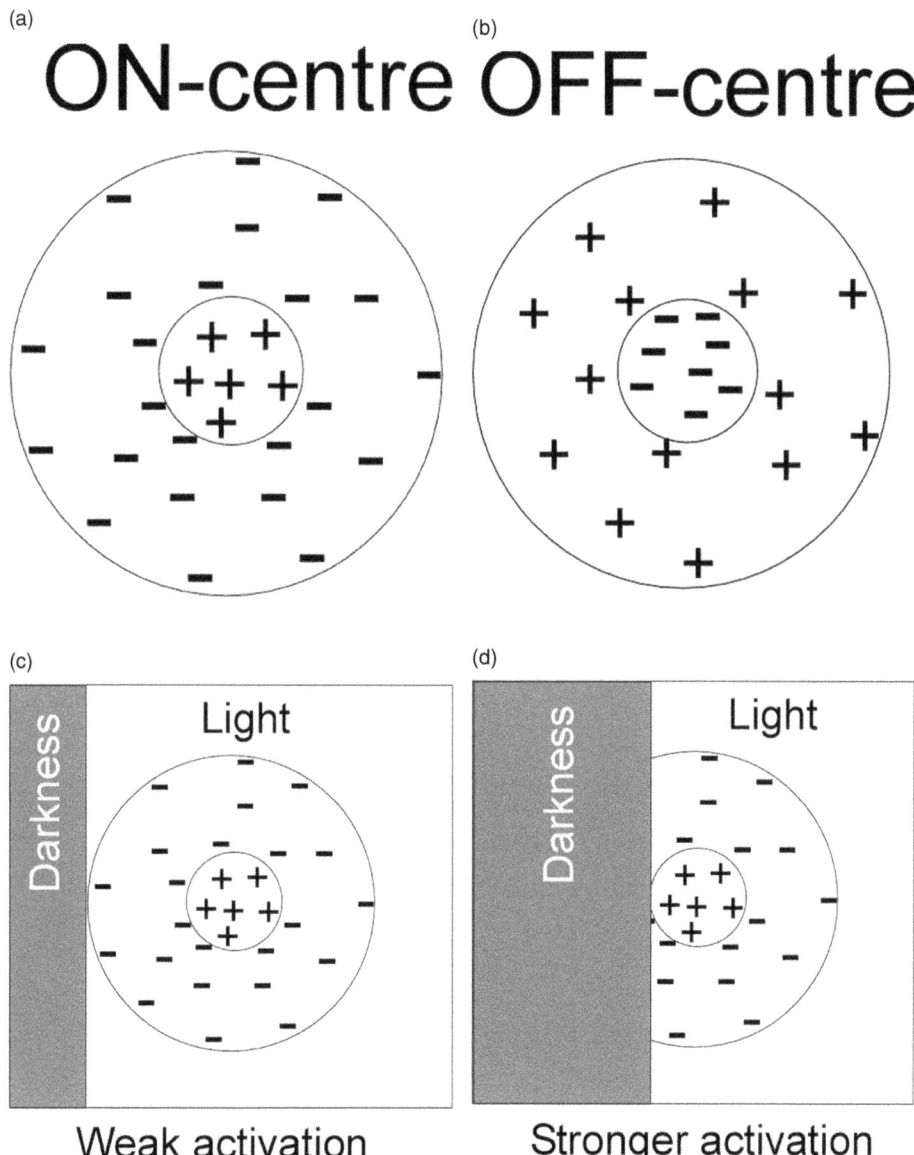

Figure 4.7 (a-b) Schemes showing how "*receptive fields*" are commonly structured for ocular ganglion cells and for visual cells in the diencephalon (LGN). The receptive field of a visual neurone corresponds to the region of the visual field (and the retina) within which a light point or pattern will affect the neuronal impulse activity (cf. Figure 4.6). For pointwise lights, the activating (+) and inhibitory (–) effects are typically localized within two concentric circular regions, either with activation in the centre and a surrouding zone of inhibition (a; *ON-centre field*) or vice versa (b; *OFF-centre field*). **(c-d)** The concentric ON/OFF structure of visual receptive fields provides the cells with a high sensitivity to the appearance of lightness contrast. Illumination of the whole field (c) gives only weak effects on cellular activity because the ON and OFF reactions will largely cancel each other (the central part of the field will typically dominate this weak response). The cellular effects will be larger for an illumination of only part of the field, covering unequal proportions of the centre and its antagonistic surround (d), i.e. in the presence of a variation of lightness within the receptive field.

Most of the ganglion cells send their axons to the LGN cells of the dien-cephalon (Plate 4.4), and also these central cells have round and concentric receptive fields, similar to those for the retinal ganglion cells. The LGN cells send their axons to the primary visual cortex (V1, Figure 4.8), and here the connections are such that the visual fields receive several different shapes and architectures, operating at different levels of complexity. 'Simple' cortical receptive fields are still such that they have opposing ON and OFF portions, being organized for discovering regions of lightness contrast. The shape of these cortical fields are, however, typically not round but linear, and their angular orientations are characteristically different for different cells. Here the first stages occur in the complex analysis of visual shapes.

Visual neurones are interested not only in lightness contrast but also in several other aspects of the visual information, such as the colours (see below) and, for instance, the movements of light patterns. Movement sensitivities may be seen for ganglion cells as well as for more central visual cells, and individual neurones might even react in different ways to movements in different directions.

4.6.5 Colour sensitivity of neurones

Some of the retinal ganglion cells are not interested in the analysis of colour; they react in the same way to test lights of widely varying wavelength composition across both the concentric portions of their receptive field. These ganglion cells, which seem to be dedicated to lightness contrast analysis, use signals from the L and M cones together. In the light-adapted eye, the light sensitivity depends mainly on the L and M cones while the S cones have no evident role here (cf. Figure 4.4 'In daylight', and Plate 4.3, curves M and L). Accordingly, most people will tend to find yellow a light colour and blue a dark one.

However, there are also many other ganglion cells which indeed give different reactions to light of different wavelengths,[116] and in such cases the cells seem to be organized for the analysis of colour contrasts between psychophysically complementary colours (Plate 4.5). Such ganglion cells might, for instance, become activated by a red light and inhibited by a green one, or vice versa. Other cells become stimulated by a yellow and inhibited by a blue light, or vice versa. In addition, detailed studies of the receptive fields of colour-sensitive neurones have shown that they are often constructed such as to become particularly sensitive to spatial colour contrast: this concerns the so-called *double-opponent cells* (Plate 4.6). Examples of such receptive fields have been found

Activation   Inhibition   Cell type

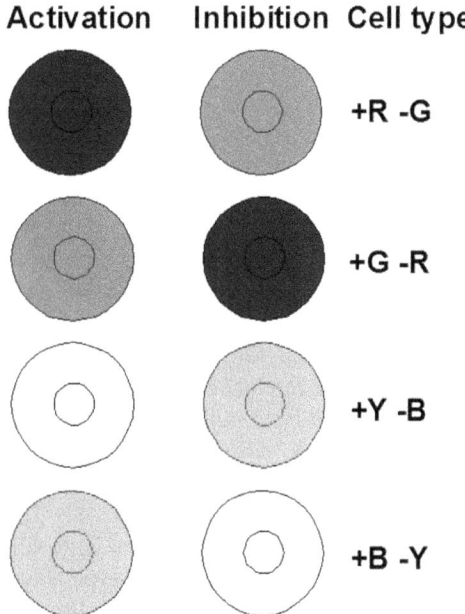

+R -G

+G -R

+Y -B

+B -Y

Plate 4.5 Many retinal ganglion cells and visual cells of LGN will, in addition to their general reactions to light distribution (Figure 4.7) also be differentially sensitive to light of different wavelength compositions (i.e. of different colour). When illuminating the whole receptive field using light of different colours, four main categories of colour-sensitive cells have been observed; (i) some cells are activated by red and inhibited by green (+R-G); (ii) some are inhibited by red and activated by green (+G-R); (iii) some are activated by yellow and inhibited by blue (+Y-B); (iv) some are inhibited by yellow and activated by blue (+B-Y). A black and white version of this figure will appear in some formats. For the colour version, please refer to the plate section.
The colours in this illustration and in the next one are purely symbolic; they are not intended to show the hues to which the various cells were reacting.

for many of the retinal ganglion cells, many of the LGN cells in the midbrain, and many of the colour-sensitive cells in the primary visual cortex V1.

The complementary reactions for red versus green seem to depend on neuronal connections which are organized such as to detect *differences* between signals from L versus M cones of the same retinal region. Similarly, complementary reactions to blue versus yellow seem to reflect a sensitivity to differences between the signals from the S cones versus the summed/combined signals from L and M cones. Such neurophysiological findings demonstrate that Hering's colour-opponent theory with its two pairs of chromatic complementary colours (red/green, yellow/blue) is quite compatible with the Young and Helmholtz trichromatic colour system: the cones have Young and Helmholtz properties (Plate 4.3) while the neuronal processing of cone-derived information leads

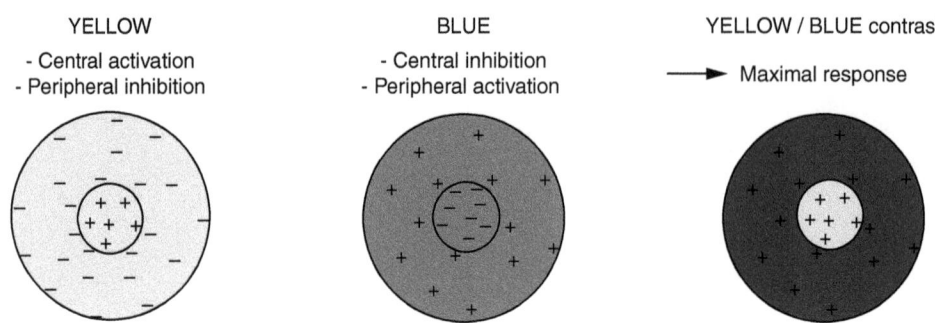

Plate 4.6 Schemes illustrating mechanisms for colour contrast. Many retinal ganglion cells and visual LGN cells and some of the visual cells in the cerebral cortex are organized such that, in addition to other properties (cf. Figure 4.7 and Plate 4.5), they react to the distribution of coloured light within their receptive field. Such "*double opponent*" types of receptive fields may be round and concentric (illustrated) or, in the cerebral cortex, they may have the shape of lines that are oriented in different directions for different cells. The colour sensitivity commonly concerns pairs of the same opponent colours as in Plate 4.5, i.e. either red vs. green or yellow vs. blue (illustrated for one yellow-ON-centre type only). Such a cell will give a maximal reaction if part of its receptive field is illuminated with one colour and the remaining part with its opponent colour, i.e. the cell will have a high sensitivity to colour contrast. A black and white version of this figure will appear in some formats. For the colour version, please refer to the plate section.

to reactions consistent with the Hering model (Plates 3.5b, 4.5, 4.6). Thanks to such neuronal calculations of *differences* between the various cone reactions, the perceptions of colour may vary largely independently of the perception of lightness. In natural scenery, the correlations would typically be high between the general luminance and the activation levels of the three individual kinds of cone, i.e. the activity levels of individual kinds of cone would not provide a sensitive indication of 'colour' (as derived from differences in wavelength distribution). However, correlations between general luminance and *differences* in activation between the various cone types tend to be very low, i.e. these difference signals provide the brain with new information.[117]

As has already been mentioned (Section 4.3.1), modern techniques make it possible to identify and count the three types of cone in living humans. Such observations have revealed that the mean numbers are different between the L, M and S cones (Section 4.3.1). For L and M cones, very different percentages were observed in different individuals; the ratio (L%/M%) varied between 1.1 and 16.5 (see note 99). Curiously, large differences in L/M-cone ratio were not associated with any measurable differences in colour vision;[118] apparently there are brain mechanisms that compensate for such numerical variations in sensor equipment.

Colour mixing can, to a certain extent, also occur between the two eyes; this provides an interesting illustration of the complex way in which the brain handles colour. For instance, use two differently coloured transparent objects (e.g. glasses, optical filters) and hold one in front of each eye. Looking with both eyes together, one might then see a new colour which arises from the intra-cerebral mixing of the separate colours seen by each eye (*binocular colour fusion*). This mixing takes place in the cerebral cortex; only here do the signals from each eye come together onto the same nerve cells. The binocular colour fusion largely occurs according to the same rules as the additive mixture of coloured lights; for instance, with an appropriate choice of wavelengths red + green may become yellow (cf. Plate 2.11).[119] The method has its limitations: subjects differ with regard to the ease with which the fusion takes place. If the binocular difference in wavelength is too great there will be no fusion but instead a *binocular colour rivalry.*[120]

4.7 The further brain analysis of colour and shape

4.7.1 Cortical regions for visual functions

Most of the visual pathway from the retina via LGN (Plate 4.4) ends up in the primary visual cortex (V1; Figure 4.8). This region has sometimes been called the 'cortical retina', because it is geometrically organized as a projection from the retina: each retinal region has a corresponding V1 region. The cortical representation of the retina occurs in relation to the visual acuity within the various parts of the visual field: central retinal regions with a great resolving power, like the fovea, are represented by a relatively much greater cortical area than is the case for more peripheral retinal regions with a lower acuity. The close relationship between the retina and the primary visual cortex is such that localized damage within V1 gives localized blindness (*scotoma*) within a corresponding region of the visual field. If the whole of V1 becomes destroyed, one becomes 'cortically blind', i.e. there is a loss of conscious visual experience. However, much of the sensory information that the brain receives is used subconsciously. This is true also for vision, e.g. for various eye-related functions (eye movements, pupil and lens control) but, also, more generally, for the visual guidance of various kinds of limb and body movements (e.g. reaching and grasping). Thus, even in a cortically blind person, after destruction of V1, the arm and hand may still be appropriately directed using visual cues which are then not consciously 'seen' ('blindsight'). Strangely, cortically blind persons may even

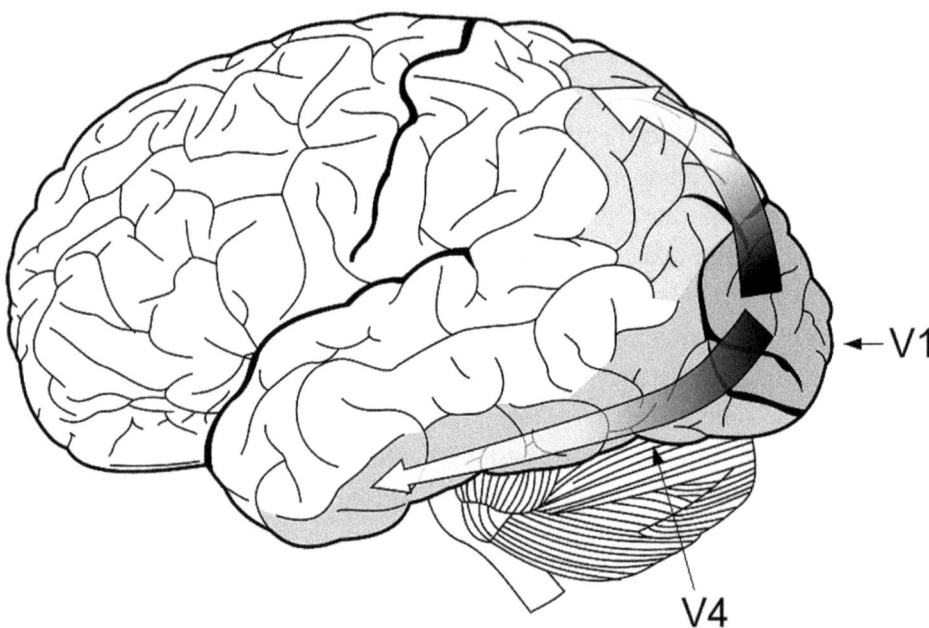

Figure 4.8 Human brain seen from the side. Most of the view concerns the cerebral cortex, *frontal lobe* to the left, *occipital lobe* to the right, *parietal lobe* in between, *temporal lobe* below. Below the temporal lobe, a small portion of the brain stem is seen and, to its right, part of the cerebellum. The *primary visual cortex (V1)* lies in the hindmost region of the occipital cortex. From V1, the visual information is transmitted to several other portions of the brain for further processing. One such region, of particular importance for colour processing, is called *V4* and its approximate position is indicated by an arrow (site on under-side of the brain).
Source: http://en.wikipedia.org/wiki/File:Ventral-dorsal_streams.svg.
Image by Selket.

possess an unconscious variety of colour vision, often being able to make correct guesses about the identity of an unseen colour in front of their eyes.[121]

From V1 there are two main projections to other cortical areas: (1) the *ventral stream*, serving functions associated with conscious visual experience; (2) the *dorsal stream*, mainly involved in subconscious 'automatic' visuomotor functions (see arrows in Figure 4.8; see note 121). The ventral stream includes the neighbouring secondary visual cortex (V2; just in front of V1) and several more frontal ventral areas. The dorsal stream gets part of its input from V1 and part via connections from the retina to the brain stem, which makes it possible to retain many of its functions also after V1 destruction.[122] The dorsal stream proceeds to various cortical areas on the upper side of the brain (including the parietal cortex).

With regard to the conscious colour vision mediated by the ventral stream, the cortical region called V4 is of particular interest; it lies further anteriorly on the underside of the hemisphere (Figure 4.8).[123] Cells within this region often react to colours, with or without combinations between the colours and other visual aspects. Bilateral damage in the region of the V4 complex can produce a state of total colour blindness: such a patient will see the world in black and white (*cerebral achromatopsia*; further information in Section 6.3.1), often combined with a lowered acuity. In addition and interestingly, damage to the V4 region may also disturb the highly complex mechanisms for colour constancy.[124] It is very unusual that a bilateral brain injury is sufficiently limited and sharply circumscribed to destroy only a small region like V4; commonly, the damage will also involve various cerebral regions in the neighbourhood. Regions close to V4 are involved in another very complex visual function: the recognition of faces. Hence, central damage causing total colour blindness is often combined with an acquired inability to recognize faces (*prosopagnosia*). Lesions to other central regions may cause the strange symptoms of *colour anomia*. Such patients may still discriminate colours from each other in a normal manner but are unable to name them; this disability may concern part of the visual field. Also on the underside of the brain lies the region V5 (also called MT), which is specialized in the processing of other visual features, including movements. Elsewhere, several other regions for various kinds of advanced visual processing exist as well. Thus, the visual information which is continuously streaming into the brain via the optic nerves is analysed in parallel at several different cortical sites for different visual features, including colour. Furthermore, as has been described in Chapter 3 (Figure 3.3), the primary information collected by the eyes, using saccadical movements, is fragmented and consists of a series of snapshots with a small sharp centre. Somewhere in the brain, all these snapshots and their colours and other special features are combined into a coherent visual image that reaches our consciousness, making us believe that this is what we are 'seeing'. It is a fascinating but still largely unsolved mystery where and how these final stages in the image processing occur.

4.7.2 Colour sensitivity and cellular organization of cortical neurones

Most of the retinal ganglion cells and the LGN cells are sensitive to both colour differences and lightness contrast, although the colour sensitivity is much better represented among the parvo- than the magnocelluar portions of the system (cf. Section 4.6.3; Plates 4.5, 4.6). The parvo- and magnocellular portions of the visual pathway are involved in both the ventral and dorsal cortical streams, but to different extents: parvo- and magnocellular components are perhaps about equally represented in the ventral stream, and the magnocellular ones show a relative dominance in the dorsal stream.[125]

Within the primary visual cortex the cells become more highly specialized, and many of these neurones are not interested in colour. Within both V1 and V2 there is an anatomical segregation between neurones processing colour and those mainly dealing with other kinds of analysis, including luminance contrast and spatial features.[126] These regional–anatomical differences are related to biochemical differences between neurones of the variously specialized locations. In the visual cortex, nerve cells with colour-selective receptive fields tend to show an intense staining for the activity of a metabolic enzyme, and these colour-processing regions are globe/spot-like in V1 (referred to as 'blobs') and stripe-shaped in V2.[127]

5 Eyes with unconventional properties: the 'red-green blinds'

Plate 5.1 shows a number of coloured dots. For people with normal colour vision, some of the dots form the digit '6', which may be seen thanks to the colour difference between these dots and the surrounding ones. However, for about 8% of all European men and and about 0.4% of the women these dots have about the same colour as many of the other ones and no numbers are seen. This picture is taken from a sensitive test for the identification of defective colour vision.

People with a deviant kind of colour vision are usually called 'colour blind', and in this chapter I will describe the most common type of this deviation, the inborn so-called 'red-green blindness', a visual constitution which is often diagnosed with test pictures like Plate 5.1. The terminology used for these varieties of colour vision is unfortunate and often leads to misunderstandings: only a very few of the so-called 'colour-blind' individuals are actually seeing the world without colour (Chapter 6), and this is certainly not the case for red-green blind people. Such persons see most or all of the reddish or greeinsh hues as chromatic colour, i.e. they are actually not at all 'red-green blind'; they do, however, have difficulties in distinguishing many of the red-green hues and nuances from each other, a property that might be labelled colour uncertainty or colour ambivalence rather than colour blindness. Unfortunately, the terms 'colour blindness' and 'red-green blindness' are so firmly established that I feel forced to use them also in this book. However, it should be stressed at the outset that, as I see it, red-green blindness should not to be regarded as a disease, but rather as one of the many ways in which people might differ from each other in their inborn constitution.[128]

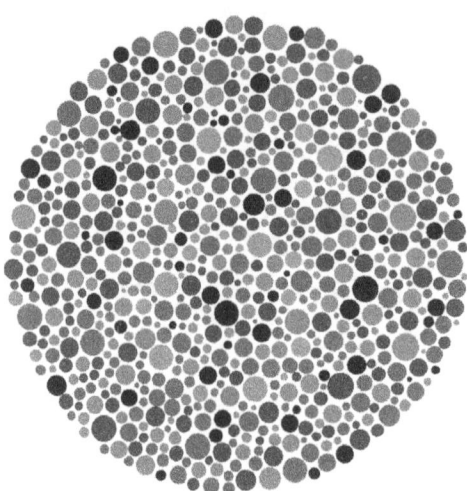

Plate 5.1 From test for colour blindness. Example of pseudoisochromatic test plate of the "*vanishing*" kind. People with a normal colour vision (normal trichromats) will see the digit "6", as formed by coloured dots of the plate. Red-green blind persons will not see any digits. From Ishihara test.
Source: http://en.wikipedia.org/wiki/File:Ishihara_11.PNG.
A black and white version of this figure will appear in some formats. For the colour version, please refer to the plate section.

5.1 John Dalton and his brother

In the world of a couple of hundred years ago, there were few (if any) colour-coded signals and the colours perceived were then a rather personal and introspective affair: people were not yet required to react to particular colours in rapid and highly specific manners. For such reasons, it took humanity a surprisingly long time to discover that people differed in their ability to detect and identify colour differences. One of the first investigators in this field was the prominent English physicist/ chemist John Dalton (Figure 5.1a). He was himself red-green blind, and in 1794 he gave a talk on his own colour vision at the Manchester Literary and Philosophical Society: 'Extraordinary facts relating to the vision of colours' (published 1798). As a child, John Dalton noticed that people in his surroundings used colour terms in a peculiar way, e.g. employing different names for the colours 'blue' and 'purple', which Dalton himself found to be very similar to each other. However, for a long time he believed this to be mainly a matter of language, 'what I conceived to be a perplexity in their nomenclature'.[129] The differences between his own colour perception and that of others became more evident for Dalton when he started with detailed studies of flowers. His most striking experience concerned the colour of a

(a)

Figure 5.1   (a) *John Dalton* (1766–1844) was a prominent physicist and chemist. He was himself red-green blind, and he published the first thorough scientific account of this constitution.
© Georgios Kollidas/123RF.com.

(b)

(b) *Lord Rayleigh (John Strutt)* (1842–1919) was a prominent physicist who got the Nobel prize in 1904 for the discovery of argon. He also discovered why the sky is blue ("*Rayleigh scattering*" of shortwave light). He made quantitative measurements of colour mixing and colour vision (e.g. the "*Rayleigh match*") and, using such methods, discovered the anomalous trichromats.
© National Portrait Gallery.

cranesbill flower, which to him looked 'sky blue' in daylight but 'red' in candle-light.[130] His brother saw it in the same way, but all other members of the household considered the flower to have the same 'pink' colour with both types of illumination. Dalton then performed a series of systematic comparisons between his colour vision and that of others, and his investigations led to the conclusion that he had a rather common inherited form of deviant colour vision, what we now call inborn red-green blindness. Dalton's conclusions were supported by the contents of a somewhat earlier publication by Huddart (1777), who briefly described the deviant kind of colour vision of a shoemaker named Harris and one of his two colour-blind brothers. The Harris brothers seem to be the very first published cases of (most probably) red-green blindness. For instance, the interviewed Harris brother 'was most of all deceived by the orange colour; of this he spoke very confidently, saying, 'This is the colour of grass; this is green.'[131] At the time of Dalton's investigations, only one of the colour-blind Harris brothers was still alive; Dalton contacted him and arrived at the conclusion that his colour vision was similar to that of Dalton himself. In his further studies of the subject, Dalton noted that his cases included no women, and that the parents of colour-blind boys were seldom colour blind themselves.

Dalton suggested that his own deviant colour vision might be due to the presence of a blue pigment in his eyes, absorbing much of, in particular, the red-green light components. After Dalton's death his eyes were kept for later investigations, which duly followed about 150 years later (Hunt *et al.* 1995). No strange pigmentation was found inside the eyes, but the DNA of its cell nuclei displayed genetic constitution characteristic for a *deuteranope*, i.e. a red-green blindness caused by the lack of functioning M cones (cf. Section 5.2).

Dalton's publication concerning the properties of his eyes had as a consequence that, in many languages, the red-green blindness was often referred to by terms like 'daltonism', which he did not mind. Thus, in French 'daltonien', in Spanish 'daltonismo', in Dutch 'daltonisme'; surprisingly, such a term is less often used in English than in several other languages. However, Dalton is most well known for his research concerning atoms and molecules;[132] his studies of colour vision represent an interesting but relatively minor sideline in his scientific career.

5.2 Kinds of red-green blindness: cones and visual pigments

In 1807, Thomas Young (Figure 2.1a) suggested that the inborn red-green blindness might depend on a lack of function in one of Young's three

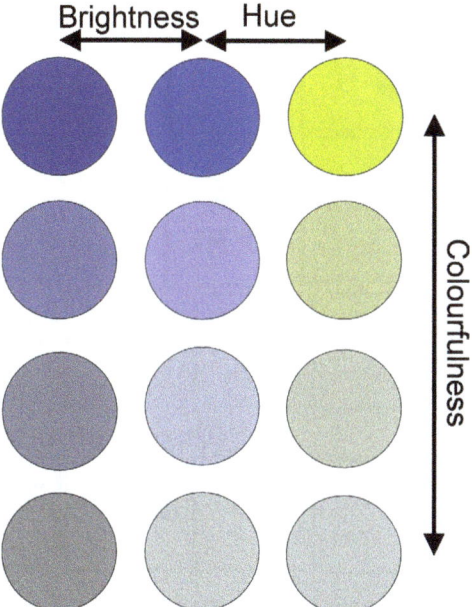

Plate 1.1

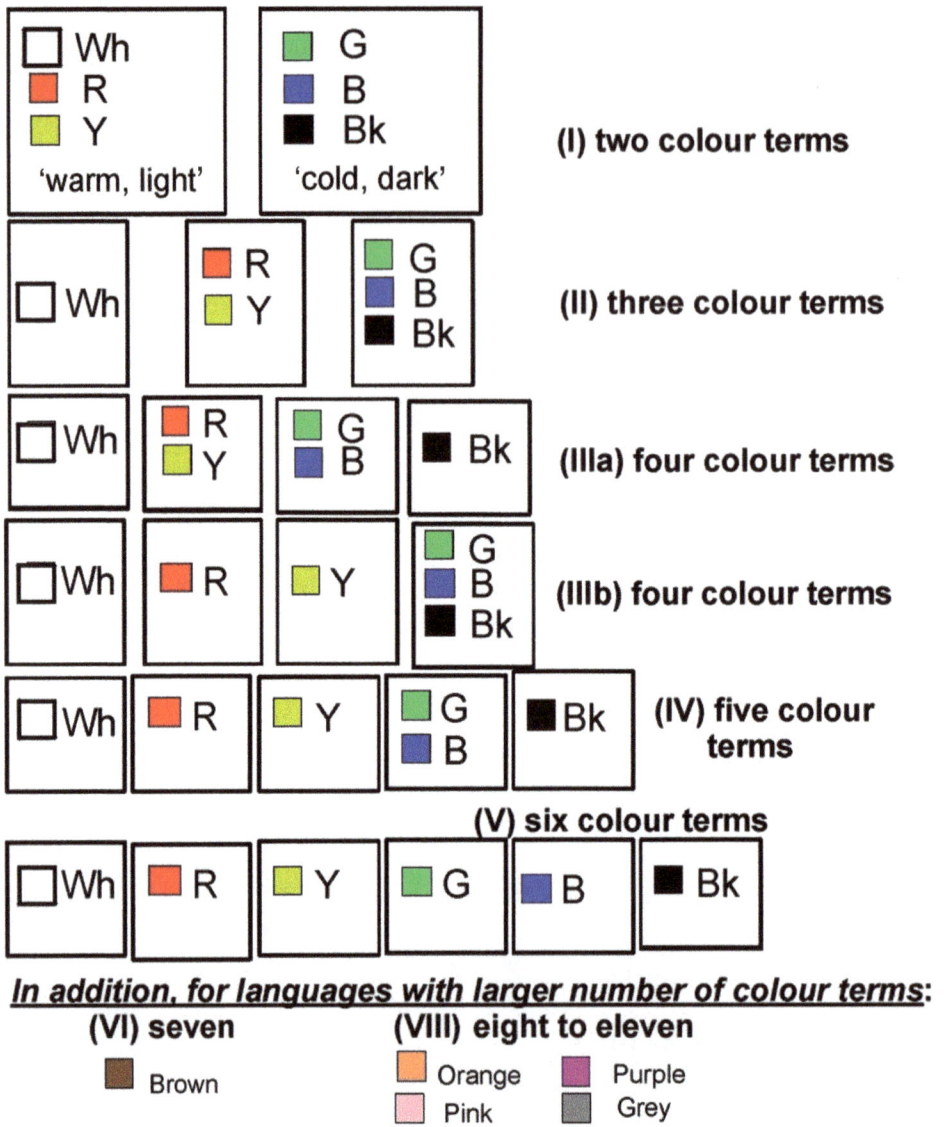

(I) two colour terms

'warm, light' 'cold, dark'

(II) three colour terms

(IIIa) four colour terms

(IIIb) four colour terms

(IV) five colour terms

(V) six colour terms

In addition, for languages with larger number of colour terms:

(VI) seven   (VIII) eight to eleven

Brown   Orange   Purple
Pink   Grey

Plate 1.2

Plate 1.3

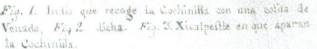

Fig. 1. Indio que recoge la Cochinilla con una colita de Venado. Fig. 2. dicha. Fig. 3. Xicalpestle en que aparan la Cochinilla.

Plate 1.4

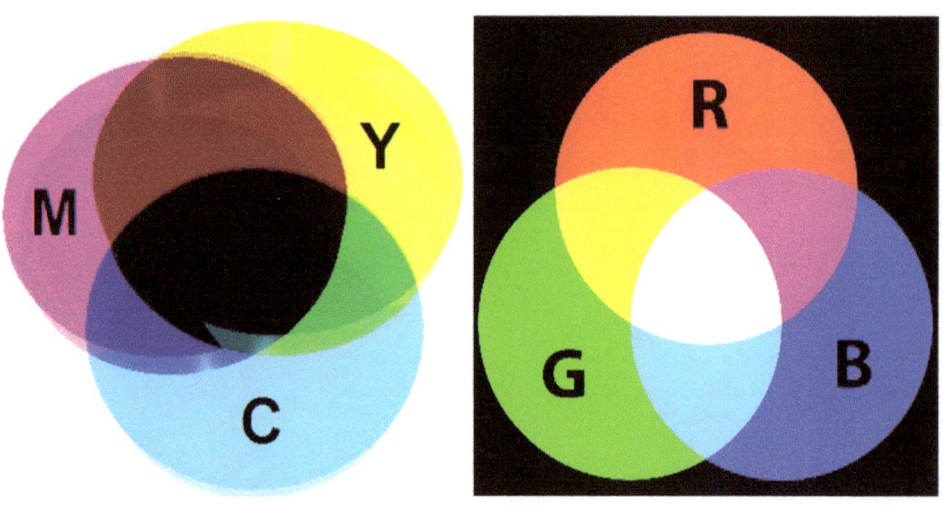

Plate 1.5

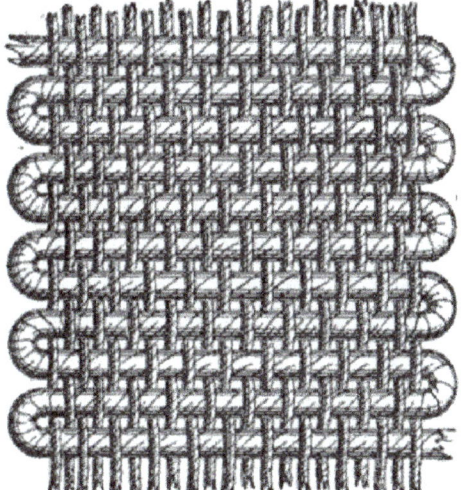

Plate 1.6

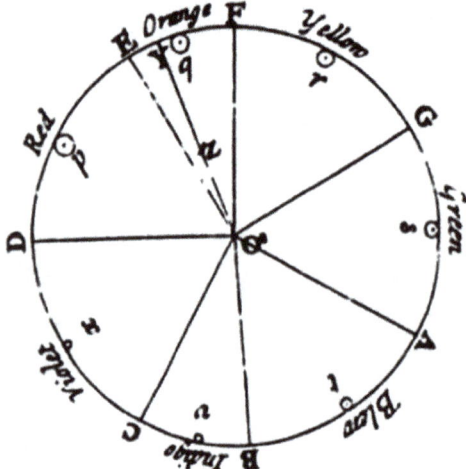

Plate 2.1

Plate 2.2

(a)

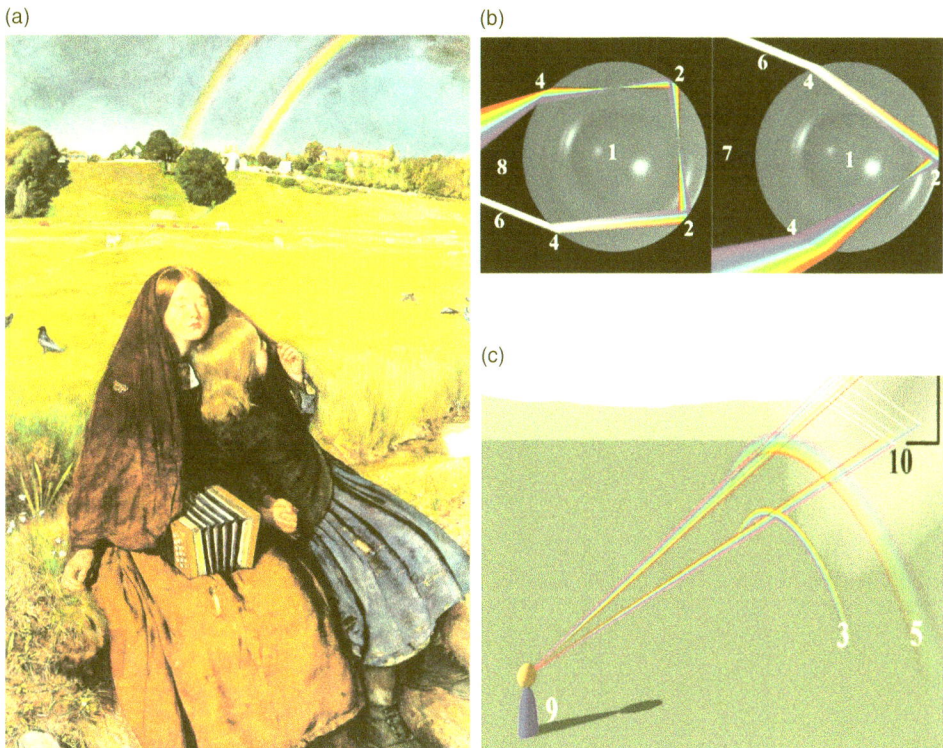

(b)

(c)

Plate 2.3

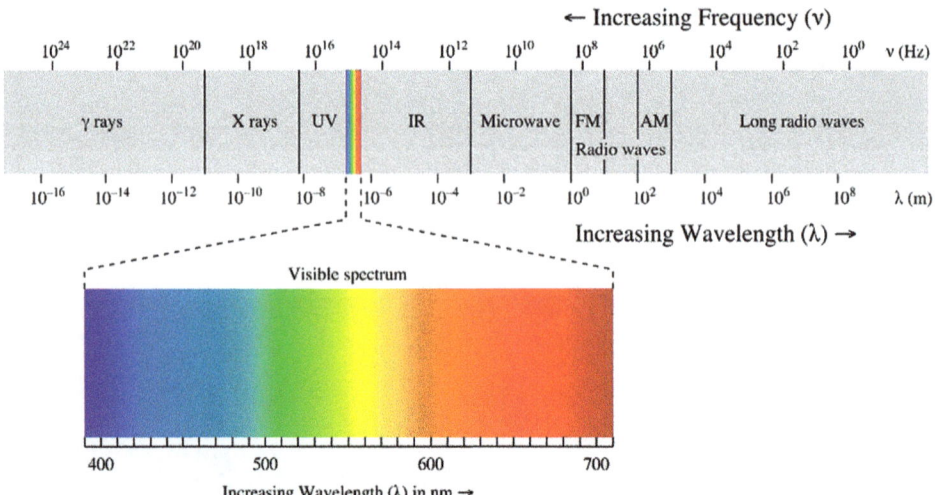

Plate 2.4

Plate 2.5

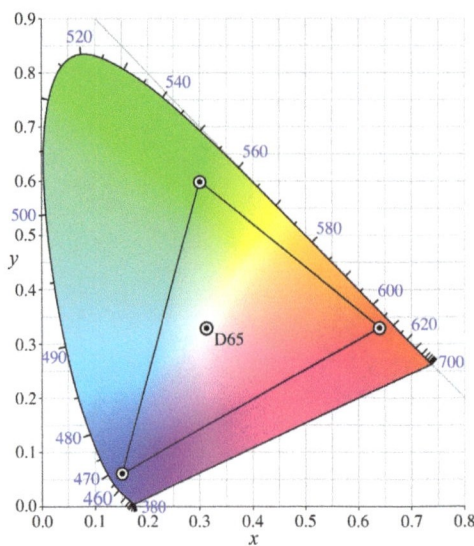

Plate 2.6

Plate 2.7

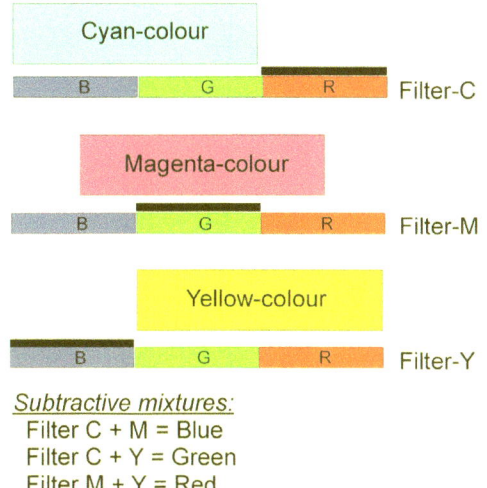

Subtractive mixtures:
Filter C + M = Blue
Filter C + Y = Green
Filter M + Y = Red

Plate 2.8

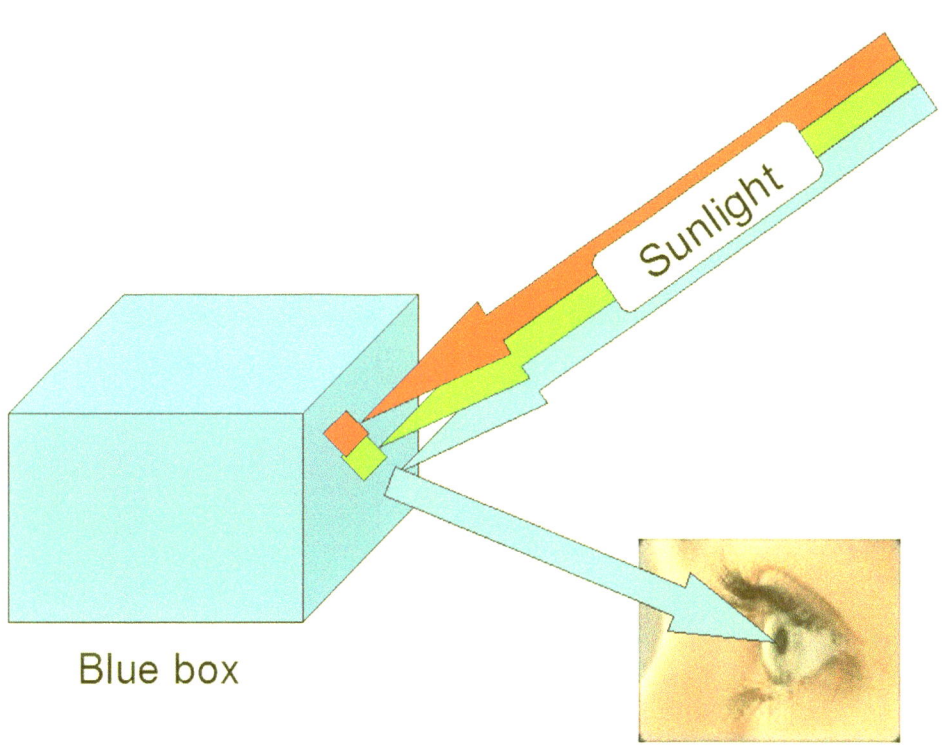

Plate 2.9

(a)

hydrogen

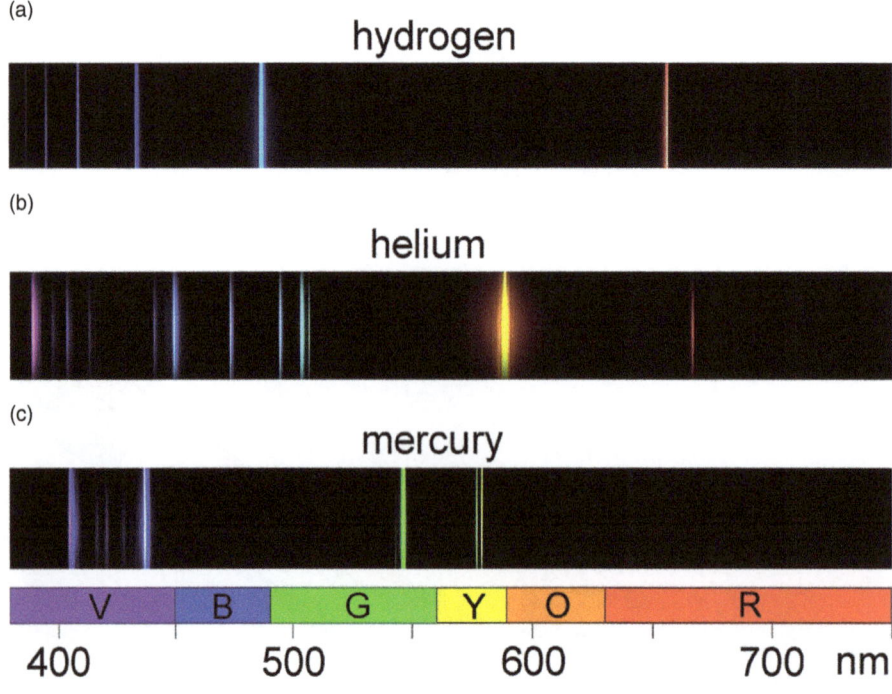

(b)

helium

(c)

mercury

| V | B | G | Y | O | R |

400   500   600   700   nm

Plate 2.10

Many different combinations of wavelengths
can produce exactly the same colour (metamerism).

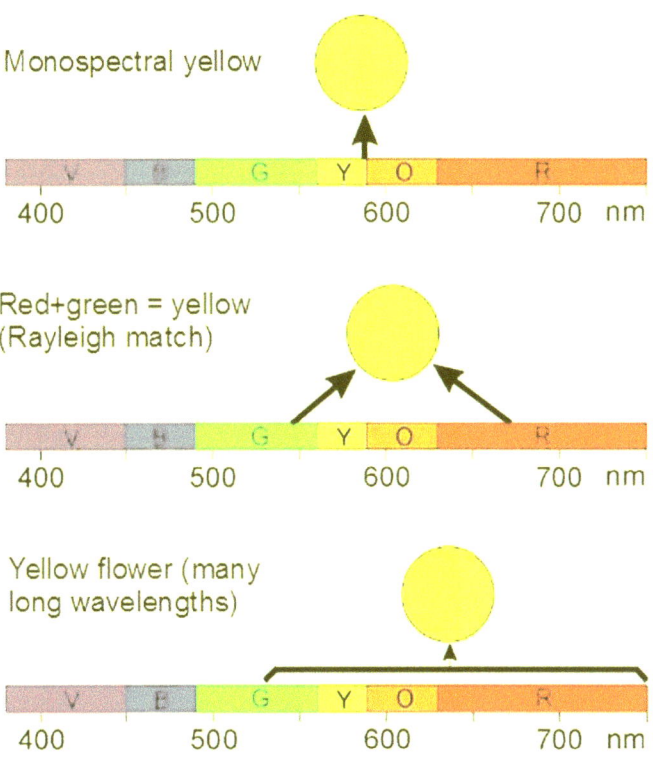

Monospectral yellow

400   500   600   700 nm

Red+green = yellow
(Rayleigh match)

400   500   600   700 nm

Yellow flower (many
long wavelengths)

400   500   600   700 nm

Plate 2.11

(a)

(b)

(c)

(d)

Chlorophyll b

Chlorophyll a

Plate 2.12

(a)

(b)

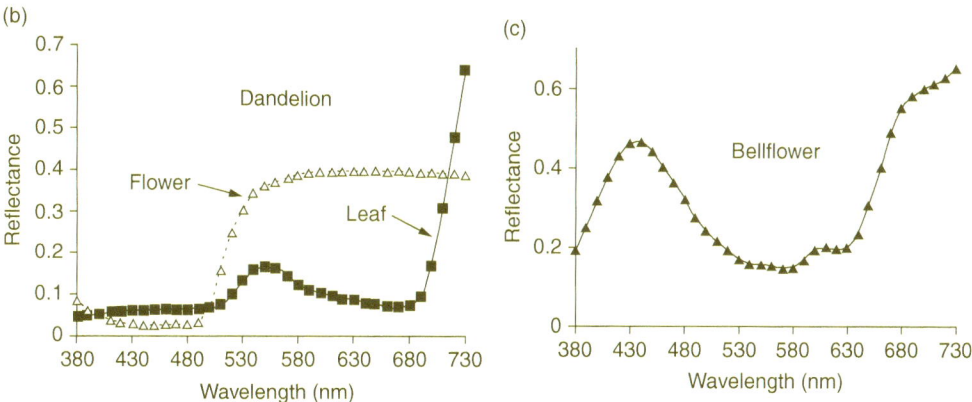

Dandelion

Flower

Leaf

Reflectance

0.7
0.6
0.5
0.4
0.3
0.2
0.1
0

380   430   480   530   580   630   680   730
Wavelength (nm)

(c)

Bellflower

Reflectance

0.6

.4

0.2

0

380   430   480   530   580   630   680   730
Wavelength (nm)

Plate 2.13

(a) Fluorescence

(b) Bioluminiscence

(c) Interference (partly)

(d) Interference

Plate 2.14

Plate 3.1

Plate 3.2

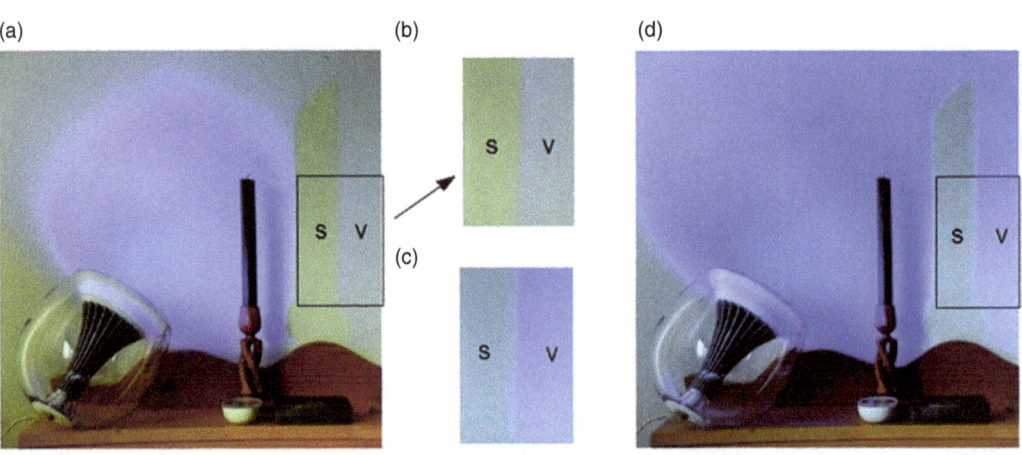

Plate 3.3

(a)

(b)

Plate 3.4

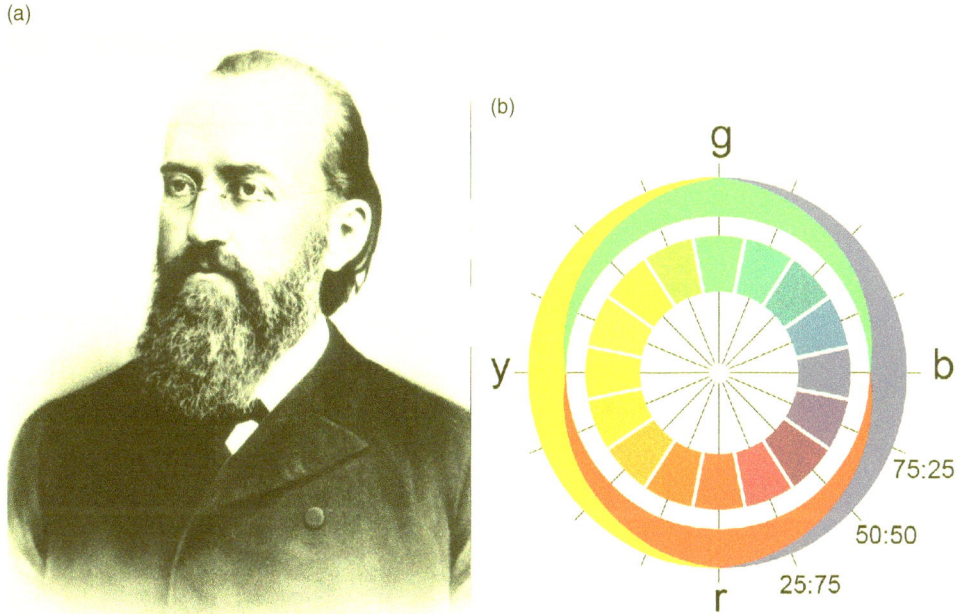

(a)

(b)

Plate 3.5

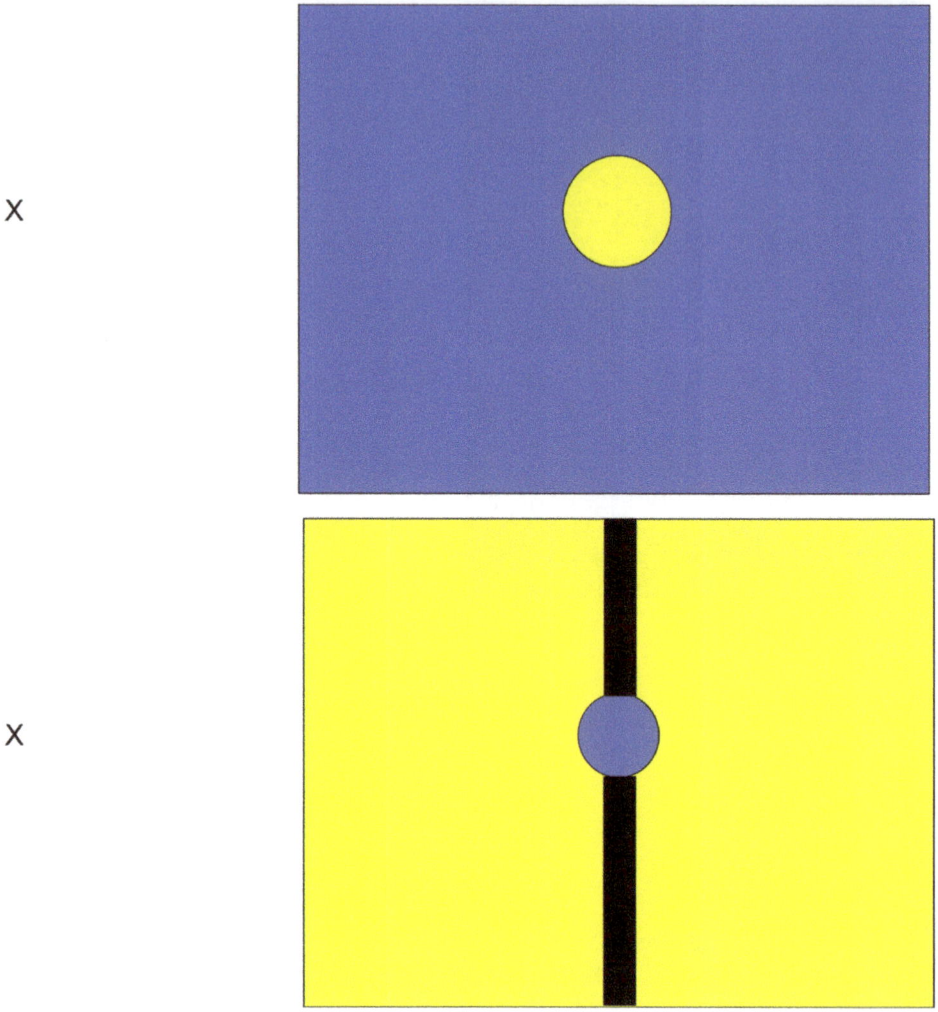

X

X

Plate 3.6

(a)

(b)

Plate 4.1

(a)

(b)

(c)

Plate 4.2

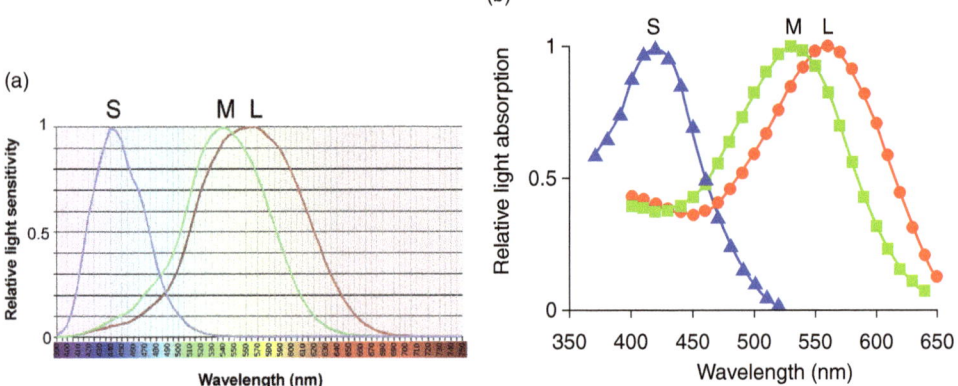

Plate 4.3

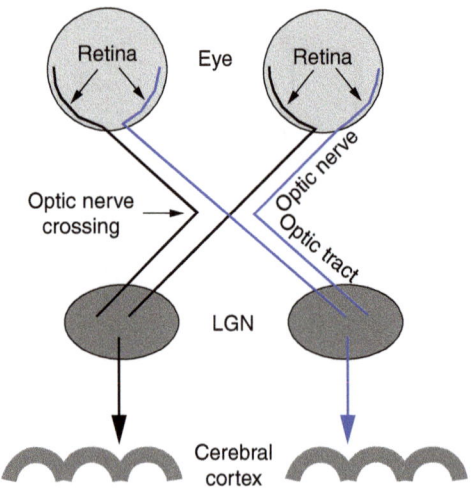

Plate 4.4

Activation   Inhibition   Cell type

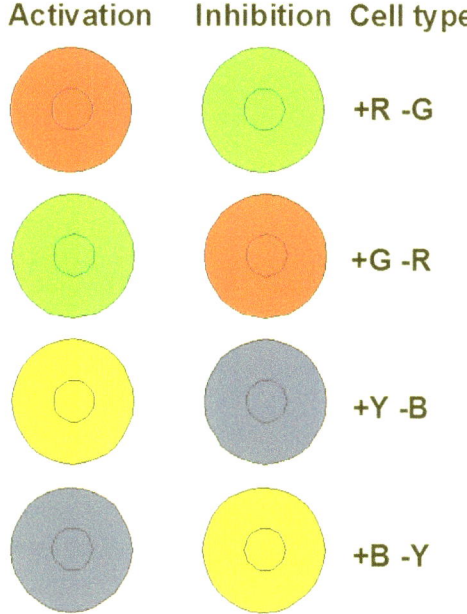

+R -G

+G -R

+Y -B

+B -Y

Plate 4.5

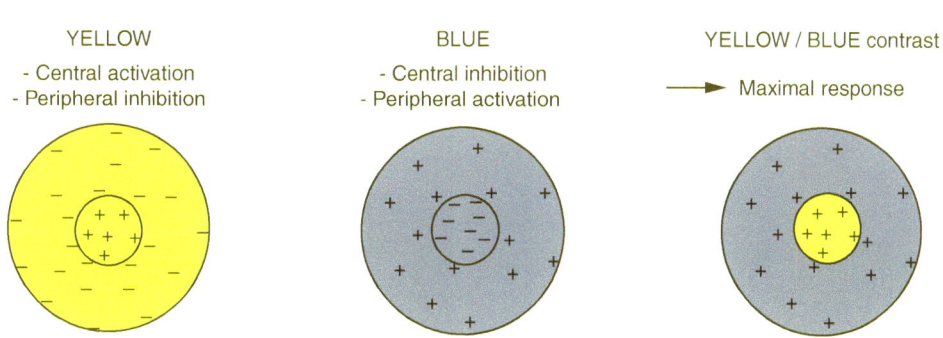

YELLOW

- Central activation
- Peripheral inhibition

BLUE

- Central inhibition
- Peripheral activation

YELLOW / BLUE contrast

Maximal response

Plate 4.6

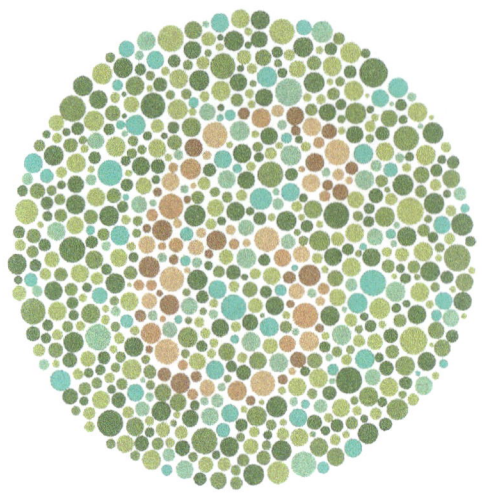

Plate 5.1

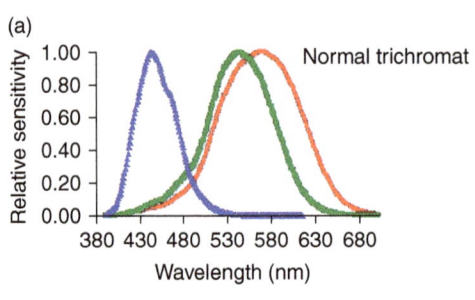

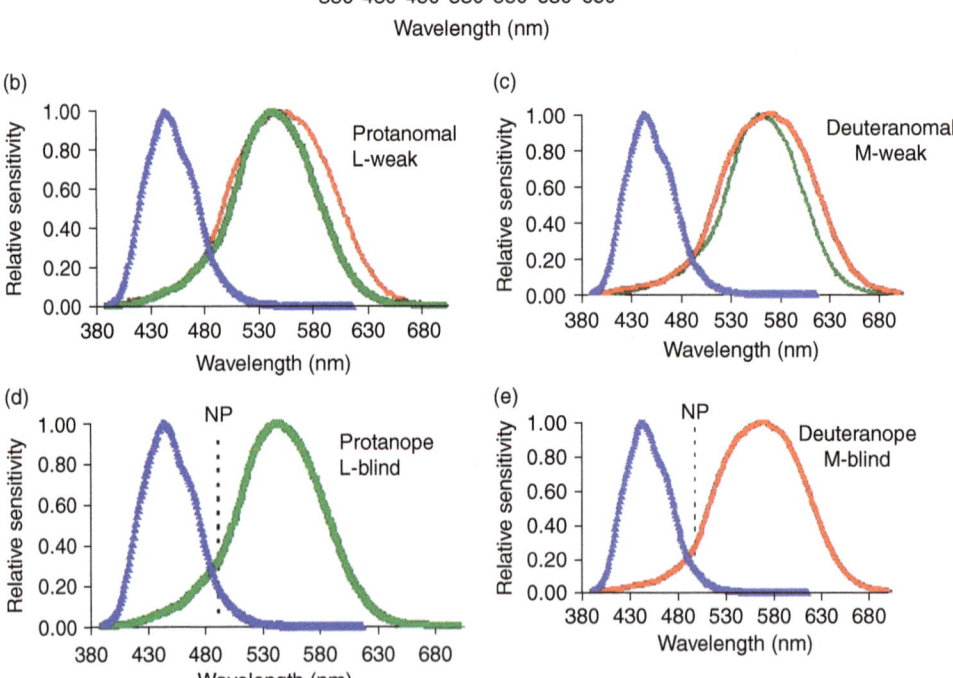

Plate 5.2

Colour analysis using filter

No filter   **Seen through red filter**

Red flower becomes light

Green label remains dark

Red label becomes light

Plate 5.3

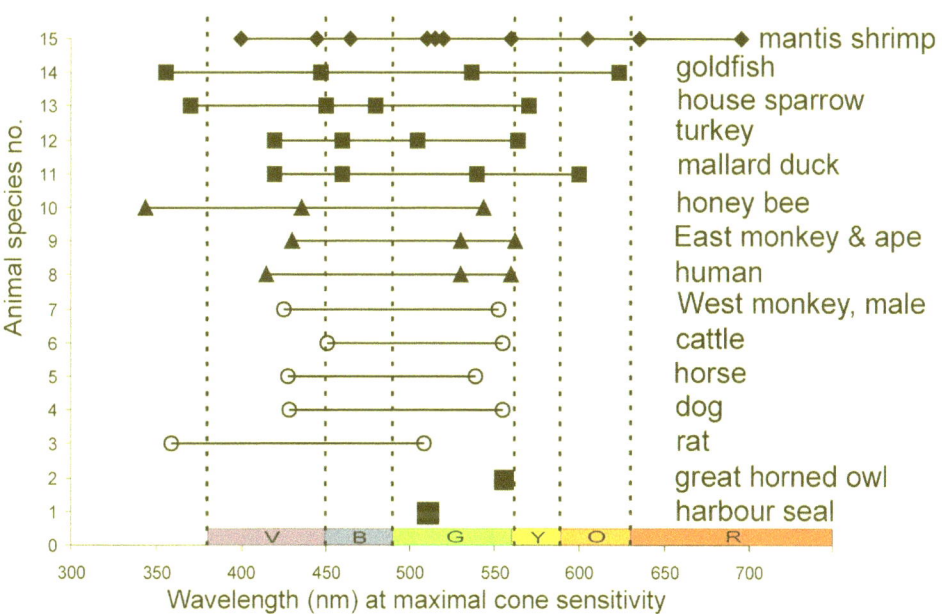

Plate 7.1

More   Similar   Less

(a)   (c)   (e)

(b)   (d)   (f)

Plate 7.2

Plate 7.3

(a)

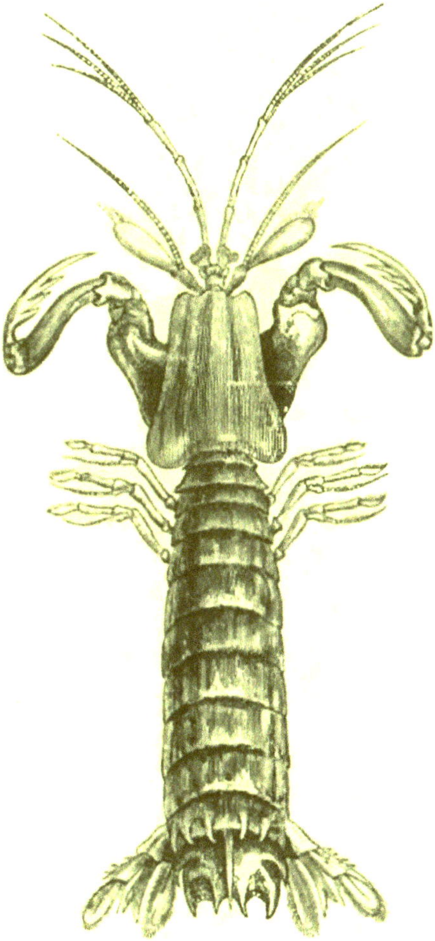

(b)

Plate 7.4

(a)

Normal colour vision

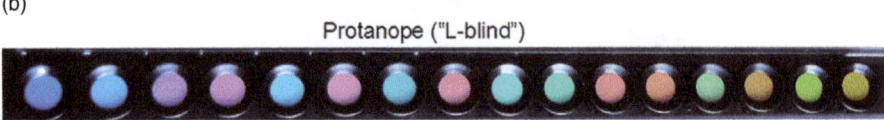

(b)

Protanope ("L-blind")

(d)

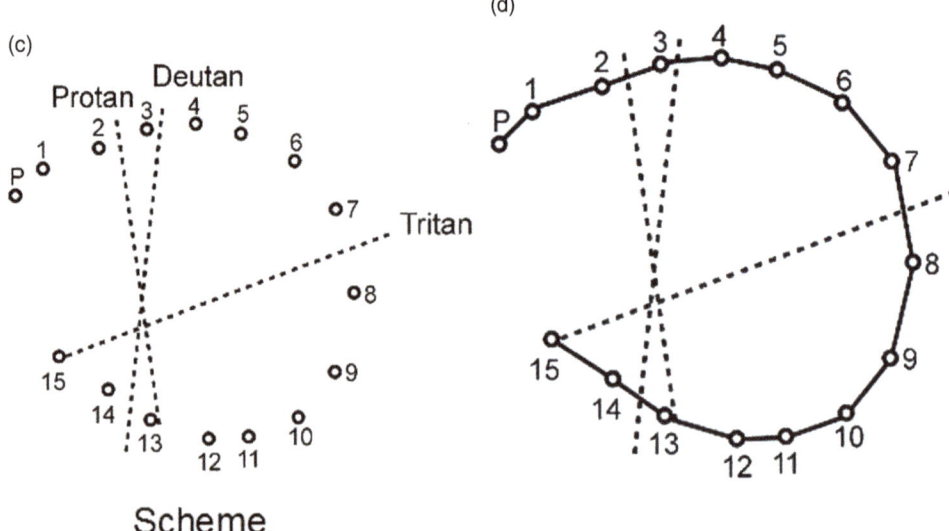

(c)

Protan

Deutan

Tritan

Scheme

Normal

(e)

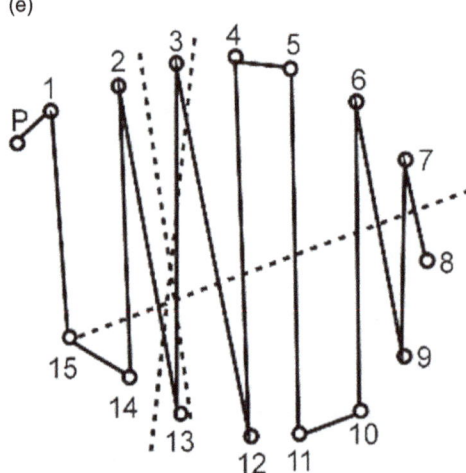

Protanope

Plate A.1

Anomaloscope

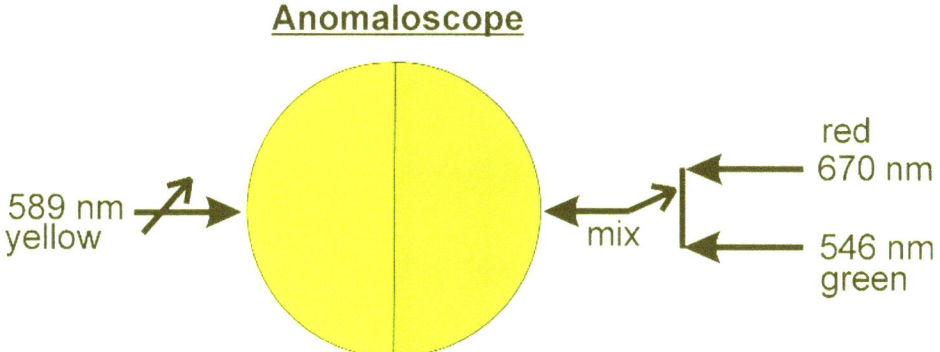

589 nm yellow

mix

red 670 nm

546 nm green

Match yellow (free choice intensity)
to mix-% red vs. green

Type colour vision	Anomaloscope-setting for match to yellow
normal trichromat	*limited variation %red matched to yellow*
anomalous trichromat	*mostly larger variation %red matched to yellow*
- deuteranomal	less %red than normal for match to yellow
- protanomal	more %red than normal for match to yellow
dichromat	*any %red matched to yellow*
- deuteranope	*c.* normal intensity yellow for match to high %red
- protanope	very low intensity yellow for match to high %red

Plate A.2

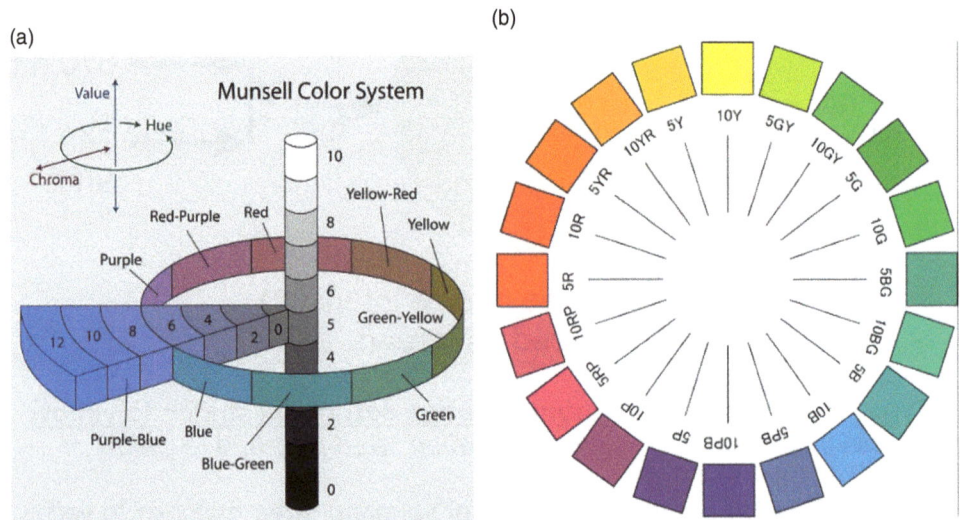

Plate B.1

(a)

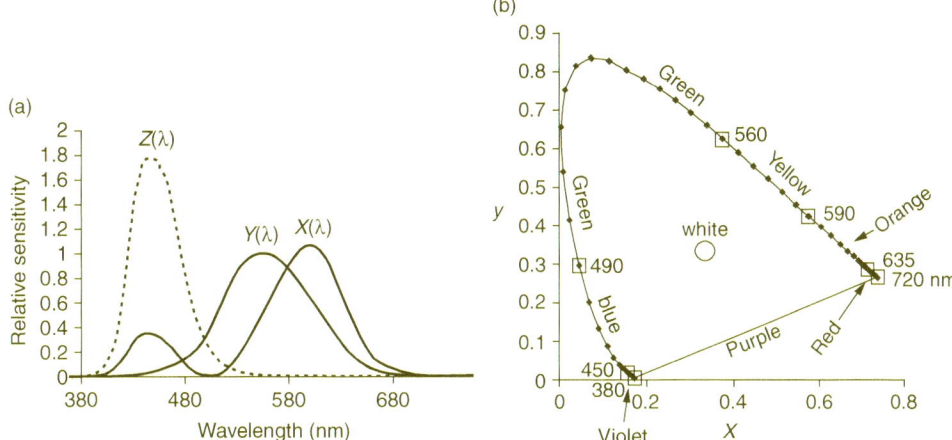

(b)

(c)

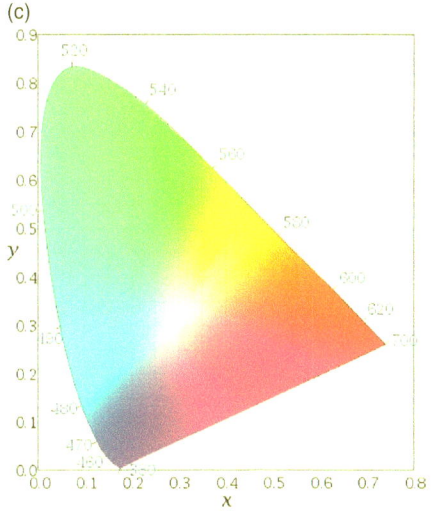

Plate B.2

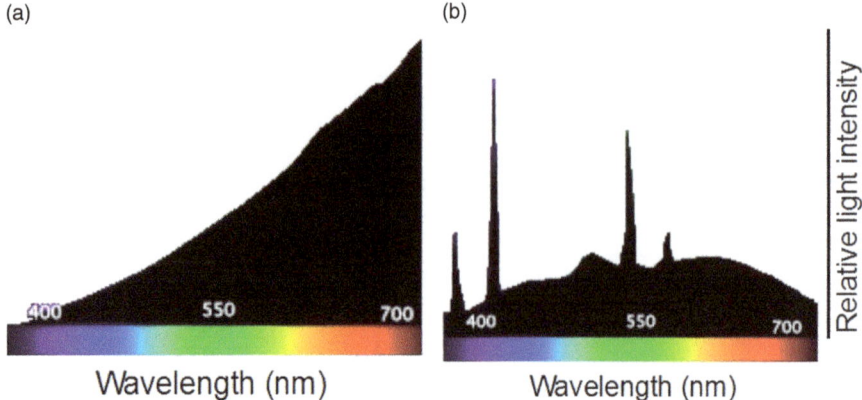

Plate C.1

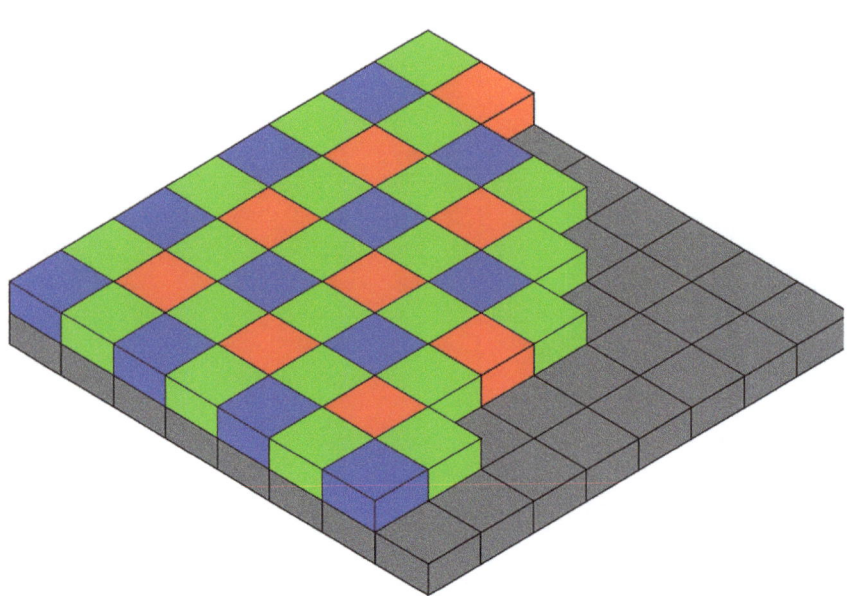

Plate D.1

Table 5.1 Occurrence of different types of inherited red-green blindness.

Gender/category	Subgroup properties	%
Male	all types	~8.0 (7.4-9.0)
Female	all types	~0.4 (0.4-0.5)
Male	*Dichromat*	2.3
Protanope	no L-cone function ('*L blind*')	1.0
Deuteranope	no M-cone function ('*M blind*')	1.3
Male	*Anomalous trichromat*	5.7
Protanomal	deviant L cones ('*L weak*')	1.1
Deuteranomal	deviant M cones ('*M weak*')	4.6

Frequencies of occurrence (%) given for each gender separately; percentages valid for Europe.
For references and further information, see Sharpe *et al.* (1999); Birch (2003).

hypothetical colour receptors; a very similar suggestion had been made even earlier by the glass merchant George Palmer.[133] Modern research has confirmed that this is indeed the case: all the inborn varieties of red-green blindness are caused by changed properties in one of the two longwave receptor cells, the L or the M cones (Plates 4.3, 5.2).

The terminology for the various kinds of red-green blindness is summarized in Table 5.1. A total lack of one class of functional cones occurs in *protanopia* (no L) or *deuteranopia* (no M); the respective persons are called *protanope* or *deuteranope*. Both these kinds of individuals are referred to as *dichromats*, i.e. subjects whose colour vision depends on only two types of wavelength-selective receptor cells (Plate 5.2d, e). These cases represent the most easily understandable variants of red-green blindness.

There are also red-green blind subjects who possess three kinds of cones, but the L or the M cones have a visual pigment with deviant light absorption properties (*anomalous trichromats*). The existence of these lighter forms of deviant red-green colour vision were discovered by Lord Rayleigh (Figure 5.1b), using his apparatus for quantitative measurements of colour mixing.[134] There are many varieties of anomalous trichromats, and their difficulties with colour discrimination are mainly caused by changes that make the light absorption characteristics more similar between the L and the M cones (Plate 5.2b, c). Such changes might concern either the L cones (*protanomaly*) or the M cones (*deuteranomaly*), and the extent of the changes varies greatly

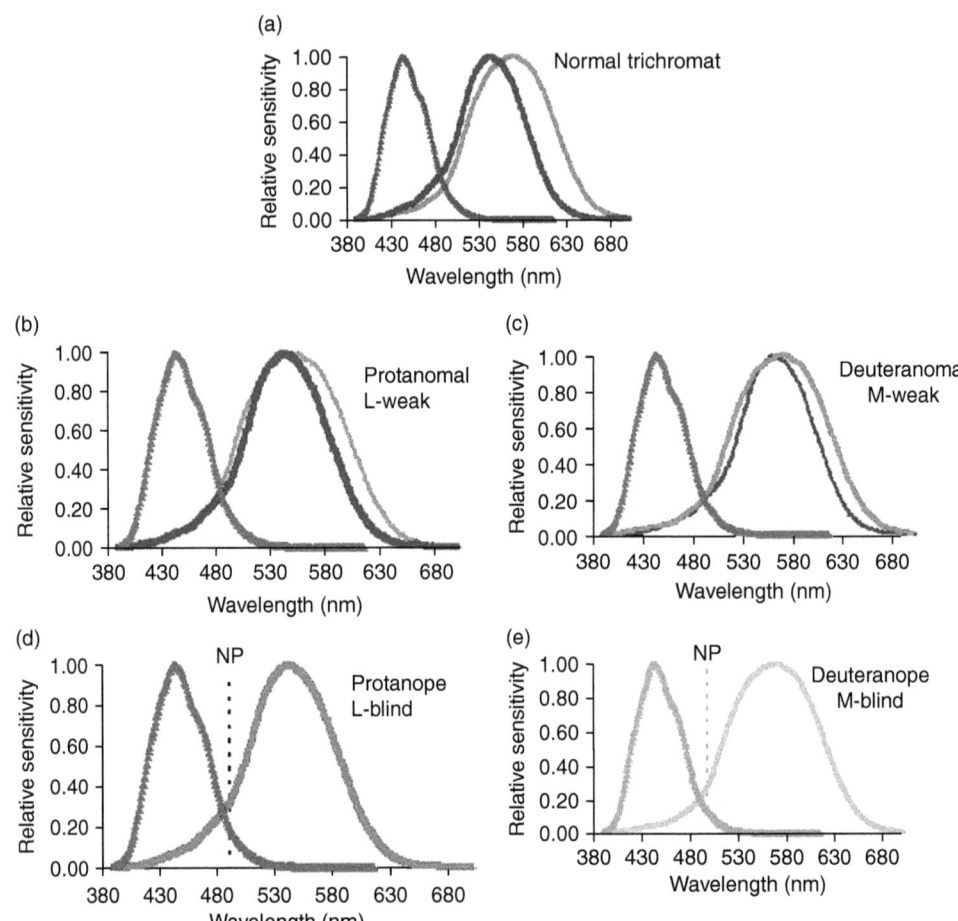

Plate 5.2 (a) Wavelength sensitivity of L, M and S cones in normal trichromats. Using data for 2 deg cone fundamentals from Stockman and Sharpe (2000), see http://www.cvrl .org/cones.htm.
(b–e) Scheme illustrating how the properties of the normally occurring visual receptor cells (a) are changed in red-green blind persons with deviant L cones (b, d) or M cones (c, e). Each graph shows the relative light sensitivity at different wavelengths for three or two different types of cone (panels (a-c): cones S, M and L; panels (d-e): cones S and M or L). (b) As in (a), but the curve for the L cones shifted toward the left (*protanomal, L-weak*). (c)   As in (a), but the curve for the M cones shifted toward the right (*deuteranomal, M-weak*). (d) As in (a), but L cones lacking (*protanope, L-blind*). (e) As in (a), but M cones lacking (*deuteranope, M-blind*). In the lower panels (d-e), a vertical interrupted line indicates the wavelength for the "*neutral point*" (*NP*; section 5.5.4). A black and white version of this figure will appear in some formats. For the colour version, please refer to the plate section.
Each panel shows the relative sensitivity of S, M and L cones (left to right) at different wavelengths of light. The normal characteristics may be downloaded from http://www.cvrl.org/ -> cone fundamentals, selecting the alternative '2-deg fundamentals based on the Stiles and Burch 10-deg CMFs (adjusted to 2-deg),' with 'Energy (linear)' (Stockman and Sharpe, 2000).

between different subjects, as determined by their respective L or M opsin genes. The wavelength for maximum light sensitivity is normally about 25–30 nm longer for the L than for the M cones (Plate 5.2a). For anomalous trichromats, this distance is often about 10 nm for protanomals and 6 nm for deuterano-mals.[135] Individual variations may be very large; in Plate 5.2b and c, the L–M peak difference is about 11 nm. In the mildest forms of anomalous trichromacy, as identified using sensitive tests,[136] the subjects might have no problems with colour discrimination in their professional and everyday life. In the most extreme cases of anomalous trichromacy, the disturbance of colour vision might be as marked as that for dichromats.

The terms 'prot' and 'deut' come from Greek words for 'first' and 'second'. This neutral terminology was suggested toward the end of the nineteenth century. Correspondingly, disturbances concerning the 'third' kind of cones (the S cones) are referred to as *tritanopia* or *tritanomaly*; such deviations are very unusual and have another type of inheritance (not sex linked; cf. Section 6.1). Some-times, colour names have been used to denote the various kinds of red-green blindness: red-blind for protanopes, red-weak for protanomals, green-blind for deuteranopes, green-weak for deuteranomals. These terms are easier to remem-ber than the Greek ones; however, I find these colour terms very unsuitable because their use might lead to serious misconceptions about how these various people see colours. For instance, protanopes and deuteranopes have very similar hue-confusion problems, and both will see chromatic colours across most of the spectrum, including its green and red portions. In this book, I will instead (partly) use another 'neutral' and hopefully easily remembered terminology which indicates the site of the disturbance: *L blind* for protanopes, *L weak* for protanomals, *M blind* for deuteranopes, *M weak* for deuteranomals (see Plate 5.2; Figure 5.3; Table 5.1).

5.3 Inheritance: from mother to son

The wavelength selectivity of the cone pigments depends on the properties of their opsins (cf. Section 4.5.1). Opsins are proteins, and the composition of each protein in the body is specified by a corresponding gene. For opsins of the two longwave cones, L and M, the genes are located close to each other on the X chromosome. As has already been mentioned above (Section 4.5.1), the L and M opsins are normally very similar to each other in their sequence of amino acids. The resemblance of their two genes is due to their rather late

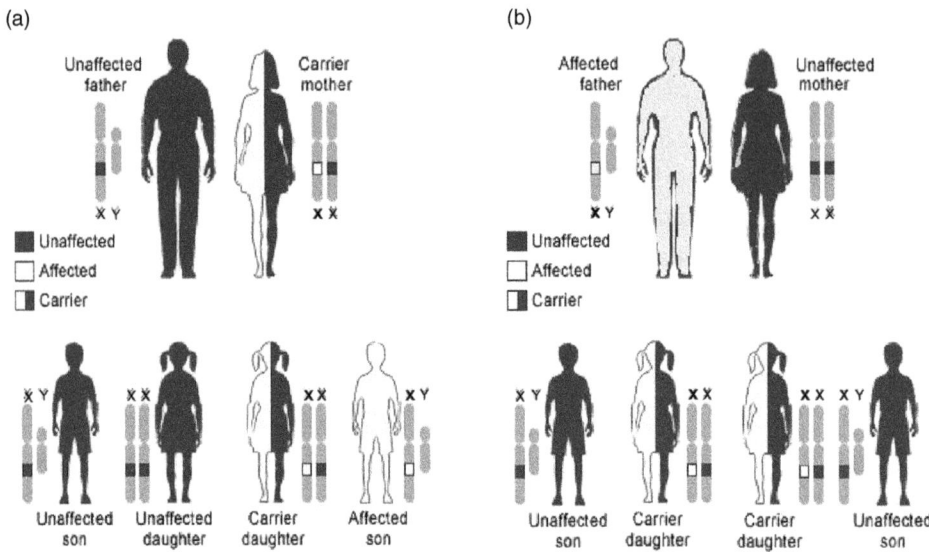

Figure 5.2 Inheritance patterns for a sex linked recessive property, such as the inborn red-green blindness. The X chromosome contains the genes for M and L opsin. Source: http://en.wikipedia.org/wiki/File:XlinkRecessive.jpg.

emergence and differentiation during pre-human evolution, at only about 40 million years ago, from a common ancestor gene (Section 7.3).

As the genes for the L and M opsins are located on the X chromosome, the functional properties related to these opsins become sex linked (Figure 5.2). In our cells, we all have 1 pair of *sex chromosomes* and 22 pairs of the other ones (*autosomes*). The sex chromosomes are of two types: a large X with many genes and a small Y with very few genes. The genes of the Y chromosome do not concern the visual pigments. A male person has one X and one Y chromosome, a female has two X chromosomes. If a female has one 'colour-blind' and one normal X chromosome, she will have largely normal colour vision, i.e. with a classical terminology the normal opsin gene is 'dominant' and the deviant one is 'recessive'. Such a female 'carrier' of a deviant L or M gene often has both normal and deviant cones in her retina, and sensitive tests might reveal some very slight deviations in her colour vision.[137] However, a female carrier has at least one normal L and M gene, producing sufficient amounts of normal L and M opsin for a (largely) normal trichromatic colour vision. On the other hand, a male subject with a 'colour-blind' X chromosome will always become colour blind, he has no other X chromosome and no backups for the L and M genes.

Children get one of their sex chromosomes from the mother (one of her two X) and one from the father (his only X or his only Y). Hence, the daughter of a red-green blind man will always inherit the deviant gene, which is present in the only X chromosome of the father (Figure 5.2). The son of a red-green blind man gets no X but only the Y chromosome from the father; hence, he will never inherit the deviant M or L genes from his father. For the children of a mother with a red-green deviant X chromosome and a father with normal colour vision, an average of 50% of the sons become red-green blind and 50% of the daughters become carriers. If such a woman has children with a red-green blind man, then also some of the daughters might become colour blind. However, this only happens if the deviant genes from mother and father are sufficiently similar. If both parents have L genes for protanopia, then 50% of their daughters become protanopes. However, if the mother is a protanopia carrier and the father is a deuteranope, then all daughters become carriers of deuteranopia and 50% also become carriers of protanopia. Intuitively surprising, such double carriers will actually have a largely normal colour vision because they still have one normal L and M gene that will provide them with normal opsins and normal visual pigments.

As already mentioned, about 50% of the sons of a female carrier of red-green blindness will themselves be red-green blind. In Western Europe, about 8% of all males are red-green blind. Hence, one might expect that about twice as many of the women are carriers. In accordance with these expectations and more exact calculations, about 15% of all women in Western Europe are carriers of red-green blindness. Only 0.4% of all women are themselves red-green blind, and only 0.02% are dichromats. Also these latter data are in accordance with what would be expected from theoretical calculations, as based on knowledge of the type of inheritance and the incidence of male red-green blindness.

5.4 Occurrence of inherited red-green blindness within different populations

Interestingly, the percentage of red-green blind persons is systematically different in different populations around the world (Table 5.2). The largest male incidence, around 7.4–9.0%, are in Europe and its traditional emigration regions like North America, Australia, etc. There are no clear differences between various indigenous European population groups. The lowest male incidence, about 2% or less, has been observed for the original inhabitants of various non-European regions. It is unclear which mechanisms might underlie these global differences in the distribution of red-green blindness. For instance, one might be

Table 5.2 Occurrence of inherited red-green blindness among males of different geographical regions.

Region (population)	%
Europe	7.4–9.0
Asia	4.2
Africa	2.6
Australia (Aborigines)	2.0
America (Indians)	1.9

Frequencies of occurrence given for indigenous populations of the various regions (Sharpe *et al.*, 1999; Birch, 2003).

tempted to believe that people who, until rather recently, lived 'close to nature' might have been in particular need of a good colour vision for their collection of various kinds of food, e.g. fruits and other plant products. Such an argument is, however, not necessarily very convincing; for instance, much of the plant-food collection was probably done by women and their incidence of red-green blindness is consistently very low.

5.5 How do red-green blind persons see colours?

5.5.1 Colours (hues) are more similar to each other

Even to a red-green blind dichromat, the world seems full of colour (one does not miss what one never knew); it is all coloured but with less colour variations than for a normal trichromat. This comes about because many colours which are different for the trichromat become fused into a single one for the dichromat. For instance, for my own protanopic eyes, there is a highly saturated and strong colour which I call 'orange-green'; to others it is either orange or green. Other examples are 'purple-blue' or 'violet-blue', all of them strongly blue for my eyes (as was also the case for Dalton; cf. Plate A.1b).

It is important to remember that red-green blind people may differ enormously in the degree of their deficiency. Those with a minor deviation might be able to identify and discriminate between all colours encountered in a normal working situation; their greatest difficulties might be to see the differences between various highly unsaturated pastel colours, a task which is seldom of great practical importance (except perhaps in clothes shops).

Measurements show how drastic the differences are in the number of hues perceived by normals and red-green blind dichromats. A normal trichromat may distinguish about 150 different hues within a solar spectrum, and differences of <1–3 nm might be large enough to give a different hue. This is very different from the number of spectral hues that may be distinguished by dichromats: about 31 for deuteranopes and only about 21 for protanopes.[138] The difference between the two types of dichromat depends mainly on the better ability of deuteranopes to discriminate between different purples. For normals, the sensitivity to changes in wavelength (i.e. the precision of hue discrimination) shows a double maximum, at about 500 and 600 nm respectively. For dichromats, differences between spectral hues are best seen within a narrow region at around 470–525 nm; within this limited range, the dichromat's sensitivity to wavelength differences might approach that for normal trichromats. Spectral colours lying at opposite sides of this sensitive region are perceived as being very different, e.g. yellow versus blue.[139]

A dichromat might believe that he/she is seeing differences in hue also *within* spectral regions above 525 nm or, somewhat less clearly, below 470 nm. However, those differences will often depend on differences in saturation and/or brightness rather than in hue. In my early life as a 'naïve' colour-blind protanope, I felt convinced that I really saw different colours in the bright yellow and in darker red portions of the spectrum. However, I now know from my own experience and experiments that my 'red' colour perception might be evoked by a very dark yellow. One should never forget that colours have three dimensions: hue, colourfulness and brightness (Plate 1.1). Therefore, even if the world of a protanope is limited to 21 hues, his number of colours is much larger.

Due to his relative scarcity of differently perceived hues, a red-green blind person (particularly a dichromat) knows and may apply a greater number of different names to colours than the number of colours that he/she can actually see. Sometimes this might lead to problems of communication. Interestingly, this might even concern achromatic colours, e.g. nuances of grey which the dichromatic person knows might have a chromatic colour for the trichromatic majority. Hence, in order not to seem too 'peculiar', a dichromat might sometimes make the wrong kind of guess and use chromatic colour names for grey or black objects because he/she is led to believe that normal trichromats might find them to be, for instance, blue-green or deep red. More often, however, problems arise in relation to the distinctions between various truly chromatic hues. To tell me that 'we will meet at the green car' in a car park, is less than ideal because

I might see a confusing number of different 'green' cars while the normal trichromat might see only one. If my companion instead points at an orange-coloured car and asks me 'what is the colour of that car?', I might equally truthfully answer 'green' and 'orange'; what I will actually be saying depends on the circumstances, perhaps I know which colour is the most likely one for this kind of car. The interviewer might get the impression that my colour vision is fine if I happen to say 'orange'. If I happen to say 'green', my eyes (or myself) are clearly mad, and the interviewer might even get the exotic idea that I see green as orange and vice versa. Wrong answers to colour questions usually lead to lots of smiles and laughter among the normal trichromats, particularly in groups of children.

Using the CIE chromaticity charts (Plate B.2), one might calculate which colours would become identical if one of the three kinds of colour sensors (cones) lost its function. In this manner, CIE diagrams may be constructed with straight lines along which the colours are lying that become identical after the loss of a particular kind of cone (Figure 5.3). For each kind of dichromacy, these 'confusion lines' fan out from a single point of origin whose position is dichromacy specific and, in addition, somewhat different for different individuals.[140] As shown by the confusion lines in Figure 5.3a and b (cf. Plate B.2), a protanope and a deuteranope have very similar difficulties of discrimination for green, yellow and orange, all of which are lying on the same line. On the other hand, blue and yellow lie on different lines and are clearly different from each other, both for protanopes and deuteranopes. For deuteranopes, but not for protanopes, several different confusion lines cross the purple connection between the red and blue-violet corners of the CIE chart (Figure 5.3a, b), i.e. as mentioned above, deuteranopes are better than protanopes in discriminating between different purples. Experiments have shown that the discrimination problems predicted by the CIE confusion lines fit well to the dichromatic reality. For red-green blinds of the category 'anomalous trichromats', the colour ambiguities occur along the same confusion lines as for corresponding kinds of dichromats, but less extensively so, covering only part of each line.

Attempts have been made to construct 'colour-blind' pictures, supposedly showing the normal trichromat how the world looks for red-green blinds (usually concerning dichromats). This has been done using the confusion lines of CIE charts (Figure 5.3), and the results certainly give an interesting demonstration of the difference in perceived colour *variation* between dichromats and normal trichromats.[141] However, the transformed picture cannot be said to show *which colours* the red-green blind person is actually perceiving, i.e. there is no

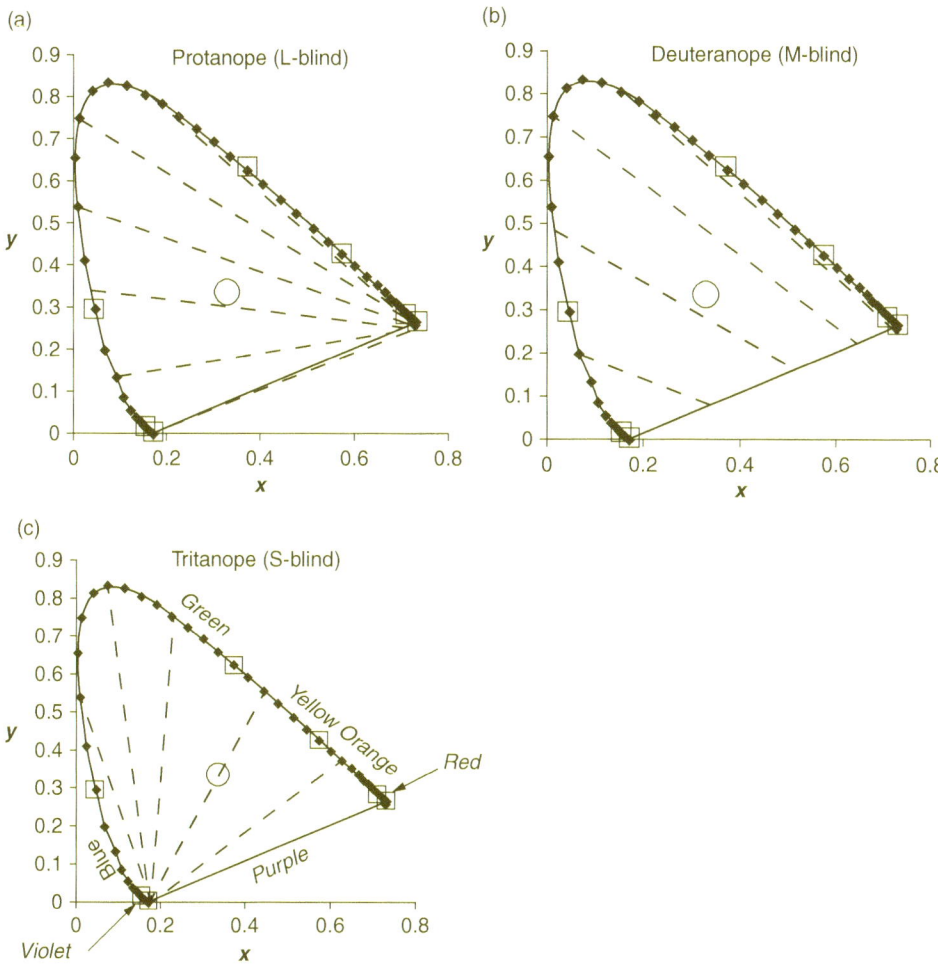

Figure 5.3 Calculated "*colour confusion lines*" (interrupted lines) for protanopes (a), deuteranopes (b) and tritanopes (c), drawn into CIE-1931 chromaticity charts. Only a selection of all the possible confusion lines are shown, demonstrating their general directions. For a dichromat, colours along the same confusion line cannot be discriminated with regard to hue. Colours become achromatic (grey/white) for confusion lines passing through the white point of the chart (central circle). One of the differences between protanopes and deuteranopes concerns their ability for discriminating between different purple hues. This is worse for protanopes and, accordingly, the CIE chart shows a nearly complete overlap between a confusion line and the purple line in protanopes (a), but not in deuteranopes (B). Colours along the outer rim of the CIE chart are approximately indicated in (c) (cf. Plate B.2 b-c).

self-evident choice of hue along each confusion line. In principle, a red-green blind person might paint hundreds of pictures of a motif with colours that are all 'realistic' for the artist himself although, with regard to its colours, each picture might look quite different for normal trichromats. None of these pictures would be better than another for demonstrating what the artist actually saw. How do you demonstrate the green-orange colour for a trichromat? It is not green, it is not orange, it is green-orange. In these contexts, evidence is sometimes cited from observations of the very uncommon cases of people with an apparently inborn red-green blindness of only one eye; in such cases, the colour-blind eye apparently sees only blue and yellow,[142] and Dalton (1798) described his own perception of the solar spectrum using only the two colour names of yellow and blue. If this is indeed the case, it is somewhat of a paradox because, in normal trichromats, the yellow colour apparently requires the simultaneous and roughly equal activation of the L and M cones, one of which is missing from red-green blind dichromats. The attempts to illustrate what colour-blinds are seeing has a long history. For instance, already at the onset of the 1800s, Goethe tried to do this in a painting, inspired by his conversations with the colour-blind Herr Gildemeister concerning perceptions of colour.

5.5.2 Colours more difficult to discover when unsaturated, when in small objects, when in thin lines

For normal trichromats, objects have to be of a certain minimum size to look coloured, and chromatic colours will become neutralized if mixed up with too much black or white (i.e. made less saturated). As compared to the normal trichromats, red-green blinds are less sensitive to chromatic colour under these various conditions. Thus, for a red-green blind dichromat, a weak rose colour might easily be seen as grey and thin red lines might easily look black. The underlying mechanisms are still unclear.[143]

For a given colour, the degree of saturation may be quantified by investigating how much white light must be added in order to make the chromatic hue disappear. Such measurements have shown that the saturation of spectral colours varies with their wavelength, both for normal trichromats and for red-green blinds (Figure 5.4). The pattern of wavelength dependence is, however, drastically different between the two groups.[144] For normal trichromats, a particularly low degree of saturation ('negative peak') is seen within a rather narrow region around 565 nm (yellow-green). For red-green blinds, the levels of saturation are lower than those for normal trichromats at practically all spectral wavelengths; the curves approach each other only at the 'negative peak' for the normals (Figure 5.4). The saturation curve also has an evident minimum for red-green blind subjects, but their 'negative peak' lies at a wavelength quite different from that for normals. A minimum of saturation is seen at around 480 nm for

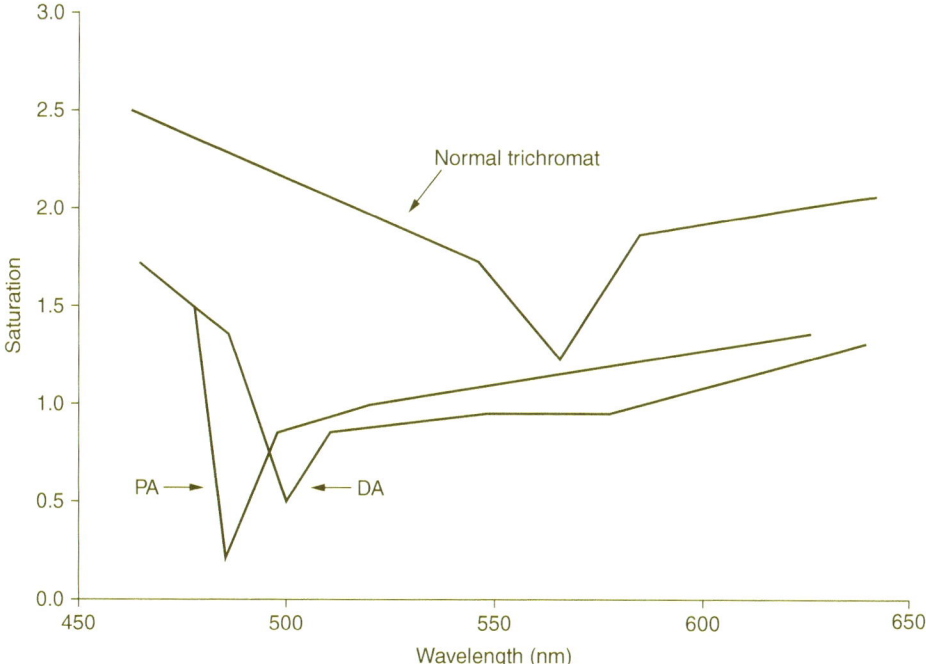

Figure 5.4 Degree of saturation for spectral colours of different wavelength, as seen by normal trichromats and by two categories of anomalous trichromats (*PA* = protanomals (L weak); *DA* = deuteranomals (M weak)). In psychophysical experiments, measurements were made as to how much white light (luminosity Lw) could be added to the light of a spectral monochromatic colour (luminosity Ls) without making it achromatic (i.e., grey/white). Values along the y-axis ("*Saturation*") show the 10-logarithm for (Lw + Ls)/Ls. Schematic representation of average data from Chapanis (1944).
Using data from Chapanis (1944).

L-deviations and at around 500 nm for M-deviations, i.e. close to the spectral transition between green and blue. For anomalous trichromats (PA and DA, Figure 5.4), the saturation still has a small positive value at this minimum; for dichromats, saturation is here zero, i.e. for dichromats this minimum corresponds to the 'neutral point' (cf. NP in Plate 5.2d, e). For such blue-green wavelengths, saturation is instead relatively high for the normal trichromats. For the red-green blinds, the regions with a low degree of saturation correspond to those with a relatively high degree of wavelength discrimination.[145]

Interesting in this context is the possible role of the rods for hue discrimination. In the central region of the fovea, only cones and no rods are present. More peripherally the rods start to appear, and they have their highest density at a short distance outside the fovea. For red-green blinds with only two types of functional cones (dichromats), hue discrimination is often better for large than for smaller coloured fields,[146] and it has been suggested that such a 'large-field trichromacy' might be partly attributed to the rods. However, under normal circumstances the usefulness of the rods as 'cone supporters' would be rather limited because they become over-stimulated and useless in normal daylight. *Mesopic* vision, i.e. vision using both cones and rods in parallel, only occurs at rather weak intensities of illumination (cf. Sections 4.4.3 and C.5), roughly corresponding to that provided by moonlight or a single candle. However, the phenomenon of large-field trichromacy has been observed also under conditions of normal illumination, i.e. rod contributions cannot be the only explanation (see note 146).

5.5.3 Red colours might seem abnormally dark (protanopes, protanomals)

For red-green blinds with L-cone deviations (protanopes, protanomals), but not for those with abnormal M cones (deuteranopes, deuteranomals), the sensitivity is lower than normal for light of the longest wavelengths: deep red colours are seen as very dark, in extreme cases turning (almost) black. This is understandable, because the L cones are the only receptors available with a sensitivity to the very longest wavelengths (cf. Plate 5.2, a versus d). The L-cone deviations might, for instance, make it difficult to see dark red numbers with a black background (a colour combination which has become deplorably common in various contexts).[147] However, most kinds of red-light signals, including traffic lights, are strong and shortwaved enough to be visible to L-deviant red-green blinds. Within a solar spectrum, the hue 'red' covers a broad range of longer wavelengths from about 630 nm up to the longest ones visible, far above 700 nm. Part of this 'red' range is covered also by the sensitivity of the M cones (Plate 5.2a).

In case of reflected light, L-deviant red-green blinds might have difficulties in perceiving any differences between black and red thin lines (e.g. letters, graphics). If lines of different colours need to be distinguished from each other by L-deviant readers, then black versus blue should be used instead of black versus red.

5.5.4 White/grey streak in blue/green part of spectrum (dichromats)

Light becomes achromatic (i.e. white/grey) if it activates all the available kinds of cones to about the same extent. For normal trichromatic eyes with cones having three different wavelength sensitivities, light of at least two different wavelengths is needed for this result (cf. Section 4.3.3; Plate 5.2a), i.e. the wavelengths of two complementary colours (Section 2.7). For an eye with only two different kinds of cones, an equal activation of both types may be produced by light of a single wavelength at about midway between the two cone sensitivities (see vertical line labelled 'NP' in Plate 5.2d, e). Thus, dichromats are truly 'colour blind' for a narrow streak in the spectrum, near the transition between green and blue. In a CIE chromaticity chart, this 'neutral point' corresponds to a confusion line passing through the neutral centre of the graph (circle in Figure 5.3a–c), and its 'dominant wavelength' is given by the point of transection between this confusion line and the chart border. The dominant wavelength of the neutral point is very similar for the two types of red-green blind dichromat, being only marginally longer for the M than for the L blind (cf. Plate 5.2d, e; Figure 5.3a, b).[148] Close to the neutral point colours are quite unsaturated (Figure 5.4), and red-green blind dichromats will see bluish-green colours as being very dull and weak. Luckily, the colours of ordinary grass and leaves correspond to more longwave spectral greens, sufficiently far from the neutral point to be strongly coloured for a dichromat (cf. Plates 2.12c, 2.13b; Figure 5.5).

Fortunately, green traffic lights often have a dominant wavelength close to that for the 'neutral point' of red-green blind dichromats. Hence, for these dichromats, a clearcut difference in colour will be perceived between such red and green lights: red/dark yellow versus light-grey/white. The International Commission on Illumination (CIE) has published two alternative specifications for green traffic lights: 'green class A' which is common and looks grey/white for dichromats, and 'green class B' which is closer to yellow and, for a red-green blind dichromat, looks more similar to amber and red lights. As a dichromat, I would be strongly in favour of having only 'green class A' at all traffic lights, particularly in countries/cities using horizontally placed signals.

5.5.5 Can red-green blind individuals see some things better than normal trichromats?

Red-green blinds see a smaller number of hues and weaker colours than is the case for normal trichromats. Hence, such people might compensate for the weaknesses of their colour perception by directing a greater portion of their 'visual attention' to differences in lightness contrast. Experimental studies have indeed shown that, as compared to normal trichromats, red-green blind dichromats are better at discovering black-and-white patterns which are covered by coloured spots.[149] With regard to the anomalous trichromats there are other intriguing peculiarities: in light-mixture experiments, their colour matching typically differs from that of the normal trichromats and they should therefore be able to see differences between mixture-colours that look the same for the normals (i.e. differences between some normally metameric colours).[150] Deuteranomalous subjects ('M weak') have indeed been found to be better than normal trichromats in discriminating between various nuances of unsaturated greens.[151] Such differences would be consistent with the interesting, but apparently never clearly confirmed, World War II rumour that red-green blind persons were then considered to be particularly useful for visual landscape scanning and for the discovery of camouflaged enemies. Compensatory talents of practical use might also have contributed to the fact that, at least for European populations, red-green blindness is still relatively common; its disadvantages were apparently not serious enough to make it disappear in the course of evolution.

A compensatory advantage for colour-blinds has also been noted among a species of American monkey, the white-faced capuchin living in Costa Rica. Male capuchins are all red-green blind dichromats whereas many of the females are trichromats and some are dichromats (see Section 7.3). Capuchin monkeys eat large numbers of insects, and the red-green blind dichromats were better than the trichromats in finding camouflaged insects at the forest floor, particularly under conditions of the deep shadows prevailing under the trees of the rain forest.[152]

It has been suggested that red-green blindness might somehow have become associated with other compensatory advantages, e.g. better vision in the dark. This thought is interesting because mammals are often particularly active at night and during dawn and dusk. Furthermore, most mammals are red-green blind dichromats; humans and monkeys are mammalian exceptions (Section 7.4). In one experimental study, dark adaptation was indeed associated with a significantly higher final light sensitivity for red-green blinds than for the normal trichromats.[153] However, in a later investigation this could not, unfortunately, be confirmed: neither for red-green blinds nor for totally colour-blind people (rod monochromats, cf. Section 6.2.1) was a greater-than-normal relative light sensitivity found after dark adaptation.[154]

5.6 Practical consequences of red-green blindness

Red-green blind people are confronted with three main types of colour-related problems:

1. irritating and/or confusing situations in everyday life;
2. colour-associated problems during various kinds of work;
3. legal restrictions with regard to professions and activities for which colour blindness is considered to be potentially dangerous.

Besides the legal restrictions, the problems encountered by colour-blind people are not widely known, largely because they tend to manifest themselves at a rather personal and private level for each individual. As compared to other groups of people with non-standard properties, there is a striking lack of organization among the red-green blinds; as far as I know, this segment of the population is hardly ever represented by any societies or socio-political pressure groups of its own. It has been pointed out that, for instance, left-handed people (c. 7–10% of the population) are much better organized. This low degree of organizational activity among the red-green blinds might partly depend on an illusion among the red-green blinds themselves: in spite of some irritating details when communicating with others, their world looks quite normal and nicely coloured, they have never seen anything else and there is probably seldom any strong subjective feeling that something essential is lacking. In addition, many of the red-green blinds have learnt not to speak too much about their deviant constitution, thereby avoiding a stream of unnecessary 'what colour is this?' questions.

5.6.1 Problems of everyday life

5.6.1.1 Irritating moments

The identification of naturally occurring colours may be of importance in several common situations, for example:

- Fruit: ripe or not? For many kinds of fruit, the ripening process is paralleled by a progressive alteration of colour, from green via yellow and orange toward red.
- Meat: fresh red or less-fresh brown?
- Growing plants: leaves healthy green or sickly yellow/brown?
- Clothes: what matches which in the eyes of surrounding normal trichromats?
- Interior decoration (furniture, carpets, curtains, pictures, etc.): same question as for the clothes.
- Waste: which bottles in receptacles for 'green' versus 'brown' glass?

Most of these kinds of problem might be neutralized, either by avoiding them, by gathering the advice of a friendly trichromat or by applying alternative and colour-independent methods. Ripening fruit often becomes darker (not only because it becomes redder), and often also softer. Clothes with potentially problematic colours might be alphanumerically labelled in various ways. And so on. The relatively manageable nature of these kinds of potential everyday problems is well illustrated by the fact that a surprising number of adults are born red-green blind without even being aware of it. This is also consistent with the circumstance that, although inborn red-green blindness is widely represented within European populations, its existence did not become known until about 200 years ago (Section 5.1).

A particular problem might concern the non-verbal communication between people. Nuances of a more or less reddish skin colour might be difficult to judge for the red-green blind, i.e. such a person might not perceive that somebody turns pale or blushes a little. For me as a protanope observer, strong blushing is evident whereas a weak blushing or paleness might escape me. As was pointed out by McIntyre in his book *Colour Blindness*,[155] such deficiencies of skin-colour sensitivity might, under very particular circumstances, possibly give others the (false) impression that the red-green blind person is a rather cold-blooded and asocial individual.

The greatest and most important colour problem of everyday life has not yet been mentioned. This is not a problem of naturally occurring colours but rather the opposite: the strong tendency of normal trichromats to use a great number of different hues for encoding important information. Unfortunately, normal trichromats have a strong preference for, in particular, red 'warning/danger' signals and green 'OK' colours. The green code lights and marks are generally not as well chosen as those of the traffic signals (Section 5.5.4), i.e. they might often be more difficult to discriminate from red for the red-green blinds. For instance, at public toilets, a red-green blind person will often be forced to use mechanical methods, trying to open the door, to find out whether the facility is occupied or not. Electrical equipment has colour-coded leads, and their status is shown by red/green LED-lamps. For the red-green blinds, such colour codes might sometimes be interpreted only with the aid of colour filters (see Section 5.6.3). Information written using dark red numbers/letters against a black background (e.g. in signs and in various kinds of electrical apparatus) might be totally unreadable for red-green blinds with L-cone deviations (protanomals, protanopes). Maps and statistical graphs are often encoded using a large number of different colour nuances, which makes this information elusive for the red-green blinds. Similarly, directions within and between various buildings might be colour coded. In financial lists, assets and debts might be distinguished from each other by the use of black or red digits, which might cause great problems for colour-blind accountants.[156] These kinds of problems have increased enormously in recent times, partly due to the great ease with which myriad different colours might be mobilized for encoding, using modern digital techniques and colour printers.

How many of the red-green blinds experience difficulties in everyday life due to their deviant colour vision? Depending on the type of situation, this was reported in an Australian study sometimes to be the case for 27–86% of the dichromats and for 11–66% of the anomalous trichromats[157]. Thus, a substantial number of red-green blinds were not aware of any problems, and surprisingly many of these adult individuals did not even know that they were red-green blind before they took part in the investigation (valid for 25% of the anomalous trichromats, but for only 5% of the dichromats). At the end of their primary school years, the level of awareness had been even lower: only 49% of the dichromats and <10% of the anomalous trichromats had then known about their deviant colour vision.

5.6.1.2 Advice for normal trichromats

When choosing colours for the encoding of information, normal trichromats must become more aware of the problems of communication between them and the large red-green blind minority. Such problems are, of course, even more serious for the much smaller group of totally colour-blind persons (Chapter 6). These kinds of communication problems might become ameliorated using various techniques, some of which have already been mentioned above:

1. Use colour-independent means, in parallel with the colours, for encoding your message. This has, for instance, been done well for vertical traffic lights: red lights above, green below, amber in between. In statistical graphics, lines might be of different thickness and/or interruption pattern. In maps, different coloured areas might also be differently hatched or dotted. Furthermore, differences in the encoding hue should be associated with differences in colour lightness; ideally, the encoding-differences should be visible also in a black-and-white copy of the map. Informative texts and arrows should be added as much as possible.
2. Avoid using only different colours for indicating the meaning of various texts and numbers (e.g. in lists of incomes versus costs). If such indications are needed, do not use red and green as encoding colours; instead use blue and black as the first choice. Yellow is, in this context, less ideal; yellow letters on white paper might be difficult to read due to their low lightness contrast. In addition, make the variously coloured letters/numbers different from each other using typographic techniques (e.g. italics, bold style, underlining, varying character size and font).
3. Avoid using colour contrasts between the symbols and their background that might make the lines of your message invisible. For instance, green-blue colours against a grey/white background might be invisible for dichromats (Plate 5.2d, e 'NP') and of very low-contrast for many anomalous trichromats

(Figure 5.4). Similarly, green letters with an orange background (or vice versa) might be invisible for many red-green blinds, and dark red colours with a black background might be invisible for L-deviant red-green blinds (protanopes, protanomals). Good advice for choices of appropriately contrasting colours are offered on several internet sites.[158]

Besides this advice concerning the choices of encoding colours, there is still another point to consider:

4. Do not unnecessarily cross-examine the colour-blind person concerning the colours in the surrounding. The results of such an interrogation are partly unpredictable because the colour-blind person has a greater number of colour names than the number of colours he/she perceives; hence, he/she might have several possible answers to each question.

5.6.1.3 Parents and schools

Together, parents and schools bear the main responsibility for helping children to adapt and cope with bewildering surroundings and other people. For problems associated with red-green blindness, much of this guidance has to be given by people with normal trichromatic colour vision. Due to the kind of sex-linked inheritance, practically all red-green blind children have parents who are normal trichromats (cf. Figure 5.2). Furthermore, particularly for lower grades, many of the school teachers are likely to be women, who are themselves rather seldom red-green blind. Thus, teachers and parents have to be informed about how red-green blind children see the world. Incidentally, the red-green blindness does not in itself impede learning at school: no statistical evidence has been found for any relationship between colour vision and school results. It is important that teachers are aware of the fact that the inborn red-green blindness cannot be influenced by some kind of 'training' and that it will not change with time or age.

General recommendations:

1. For a proper guidance of red-green blind children, it is important that they are identified as such. Thus, colour vision should be investigated for all school children, using both a sensitive test to discover colour blindness and, for those identified as colour blind, another test to indicate the seriousness of the deviation.[159]
2. In schools, teachers should give some individual advice for each colour-blind child; on average, this concerns about 1–2 children for each class of 30. The

children should be informed about what their deviation means and how misunderstandings concerning colours might arise between them and other children. In addition, when they reach one of the higher grades the red-green blind children should be told about how they might learn to compensate for part of their deviation using simple methods, e.g. employing optical red filters as an aid for the identification of some of the common encoding colours (see Section 5.6.3). Ideally, red-green blind children with a moderate/strong deviation should get an appropriate optical filter free from school, including a brief user's manual.

5.6.1.4 Car driving

All countries use red and green traffic signals. In spite of this, investigations have generally shown that the risk of traffic accidents is not higher for red-green blinds than for normal trichromats.[160] I have, myself, driven cars in many different countries around the world, and I have never found that that my protanopic eyes gave me any particular problems during such activity. Such an absence of serious traffic-associated problems depends on several different factors (partly already mentioned above):

1. Not only the colours but also the positions of the various lights give a clue as to their meaning (e.g. in vertical signals: red for STOP above, green for GO below).
2. The colour of the green lights is generally selected such that, for a red-green blind dichromat, its colour is clearly different from red (e.g. green seen as white/grey, red as dark yellow; Section 5.5.4).
3. For some kinds of traffic light, additional secondary clues have been added for distinguishing between the red STOP and the green GO signals. For pedestrians, standing-versus walking-person symbols might be shown (occasionally even animated). Sometimes, STOP versus WALK are instead shown as texts. The red and green lights might have different shapes and sizes. One of the lights might flicker, or they might both flicker but at different rates.

In modern society and, particularly, in thinly populated portions of a country (like, for instance, much of rural Sweden), it is a serious handicap not to be able/allowed to use a car. In 1956, the World Health Organization (WHO) of the United Nations issued a general recommendation that colour-blind persons should be allowed to drive, without restrictions.[161] Fortunately, most countries act accordingly and allow red-green blind persons to get a driving licence.

Exceptions still exist in a small number or countries, such as Romania and, to some extent, Japan.[162]

5.6.2 Problems at work

Colour plays an important role in many kinds of occupation. A colour-blind person should be aware of the problems that he/she might encounter in various kinds of work. This concerns, for instance, some of the activities associated with agriculture, gardening, geology and mineralogy, medicine and dentistry, chemistry, microbiology, electronics, textile and graphic design, interior decoration, photography, painting, etc. However, a red-green blind person might pursue a successful career within these various kinds of occupation, provided that he/she specializes in directions for which colour is less important and/or that he/she is aware of his/her handicap and asks advice from a normal trichromat when needed.

Red-green blindness seems to be as common among pupils of art schools as for the rest of the population.[163] It is uncertain how many and which of the famous painters of the past might have been more or less colour blind. Names that have been tentatively mentioned, but not confirmed, include Uccello, Constable, Whistler and Mondrian. The very few certified cases include the French cubist Fernand Léger (1881–1955)[164] and the Australian artist Charles Nuttall (1872–1934; best known for a large monochromatic painting).

Due to potential safety risks involving many people/passengers, red-green blind individuals are not allowed to work professionally as sea captains, train drivers or aircraft pilots. In addition, red-green blindness might, in some countries, be considered unacceptable for careers in the armed forces, as a policeman or firefighter. According to an Australian study, red-green blindness caused an undesirable restriction of the choice of vocation/employment for 43% of dichromats and 29% of other red-green blinds.[165] In many cases, this concerned occupations from which colour blinds were legally excluded due to safety issues. With regard to the details of such legal restrictions, different countries have different rules and the information is not easily localized on the internet (each country should have its own information site on such questions). In many countries, red-green blind persons may not fly any kind of aircraft. However, in some countries, individuals might acquire permission to fly small private planes; e.g. in England for flying in daytime. For some countries and occupations, a distinction is made between light and more severe versions of red-green blindness. For instance, persons might in such cases be judged acceptable if their kind of red-green blindness is mild enough to pass the D15-test (Plate A.1); this is, for instance, the case for the Canadian army.

With regard to possible occupations and careers for red-green blinds, the absolute and legal restrictions are almost completely due to the unfortunate choice, long ago, to use green versus red railway traffic signals instead of, for instance, yellow versus blue. At the time when these problems were first considered in detail in relation to the many red-green blind members of the population, the techniques were not yet available for strong blue/yellow signals. Swedish physiologist Frithiof Holmgren (1831–1897) discussed these problems in a publication of 1877, and he pointed out that 'the flame of the lamp, using turnip oil and kerosene, like all our common artificial sources of light, emits only rather small amounts of blue light';[166] he concluded that, for these reasons, a blue filter in front of the signal lamp would make its light almost invisible from far away. Old-fashioned oil lamps did indeed produce much more long-wave than shortwave light and, hence, their light intensity was decreased less by red or green filters than by blue ones. Nowadays, using modern electrical light sources, there would be no technical problems in producing blue signal lights of sufficient intensity for all kinds of traffic. However, the red-green choice has become so firmly established all over the world that a change is highly unlikely. Thus, due to the deficient capacities of ancient oil lamps, hundreds of thousands of people remain excluded from several fascinating fields of occupation. As a child at the age of about 10–12 years, I dreamed about a career as a sea captain but at the same time I sadly realized that, due to my red-green blind eyes, this would not be possible. Perhaps luckily so, for I have a strong tendency to become seasick.

It is often said that red light has a superior ability to penetrate rain and fog, which might make it particularly suitable as a warning signal. However, this is a myth: all the wavelengths of light pass equally well through fog (or rain). In fog, the light scattering is the same for all wavelengths, and this is why fog and clouds are grey or white. The blueness of a clear sky appears, of course, because the air causes light of short wavelengths to become more scattered than light with longer waves. However, such a wavelength-dependent *Rayleigh scattering* only happens for collisions between rays of light and particles smaller than the light's wavelengths, e.g. the oxygen and nitrogen molecules of air. The water droplets of fog and clouds are, in comparison, of gigantic sizes.

For light signals, colour discrimination becomes a more difficult task if one has to react to only a single colour, without any comparisons or secondary clues.

Such situations might appear in relation to aircraft traffic and, in particular, at sea when judging various orientation lights, e.g. the different coloured sectors in signals from lighthouses. However, both in the air and at sea, the importance of coloured lights has gradually waned and most of the orientation now occurs using signals from radio beacons (Decca, until the year 2000) and satellites (Global Positioning System, GPS). For professional air and sea traffic, colour vision is nowadays probably mainly of importance in cases of electronic malfunction, when light signals might have to be used as a backup mechanism.

As early as 1855, testing of the colour vision of (potential) employees at the French railways began, and legal restrictions concerning the kinds of employment allowed for colour-blinds have a long history. The first strict rules were introduced after a series of ship and railway accidents. A well-known case, much discussed at the time, concerned a train collision at Lagerlunda in Sweden in 1875. It was suggested that red/green light signals might have been misjudged by the engine driver, perhaps due to colour blindness. However, this was all speculation; the driver died in the accident, his colour vision had never been tested and there was no strong evidence that light signals played an important role.[167] It is even uncertain whether colour blindness has ever been an important (contributing) cause for a train accident. Nevertheless, the Lagerlunda collision inspired Frithiof Holmgren to develop new methods for the systematic testing of colour vision (see Appendix A).[168]

5.6.3 The usefulness of colour filters and other aids

At the onset of World War I, many young men in England wanted to enlist as fighter pilots. This task required normal colour vision, and thus about 8% of the young men had a problem. At the time, this led to lots of badly informed attempts to 'cure' the red-green blindness, e.g. by training, electrical stimulation, injections of iodine or snake poison, vitamins, etc. Nothing helped and there is still no useful treatment to cure an inborn red-green blindness. However, this does not mean that nothing can be done:

- In situations requiring the identification of various code-colours, optical colour filters might provide useful support (see detailed description below).
- Smart phones and digital tablets may be provided with apps that use the camera of the device for the identification of colours.[169]
- Furthermore, in the (how near?) future new molecular methods might become available for treatment (see Section 5.7).

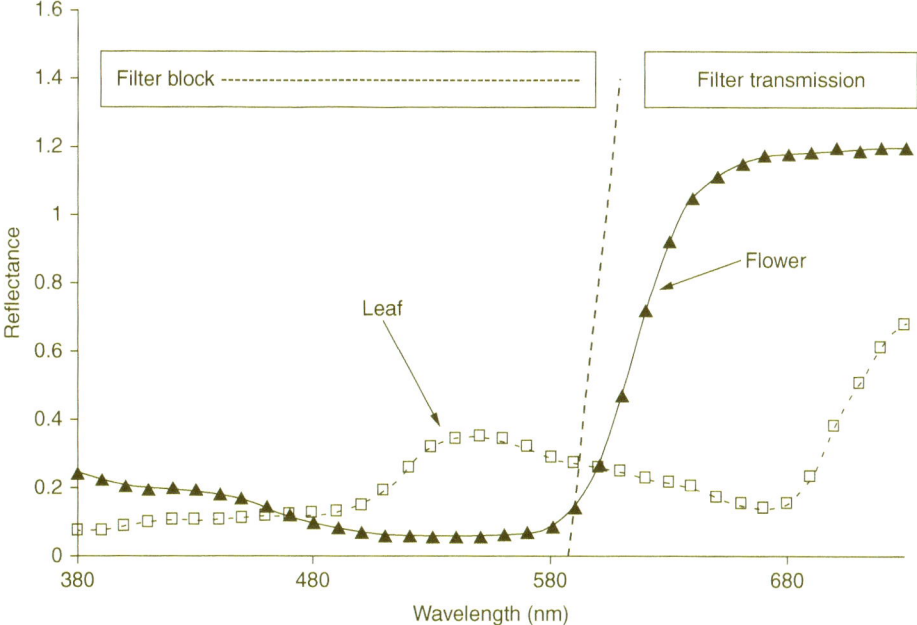

Figure 5.5 Spectrographic analysis of the experiment of Plate 5.3. Plot showing the intensity of different wavelengths of the light as reflected from a cactus flower (triangles) and a cactus leaf (open squares) when illuminated using white light. The wavelengths transmitted by the red filter of Plate 5.3 are indicated in the chart (no filter used in the spectrographic measurements). Much of the reflectance of the flower falls within the wavelengths transmitted by the filter, and much of the reflectance of the green leaves falls in the blocked wavelengths. Therefore, as seen through the red filter, the flower becomes relatively light and the leaves darker.

An optical filter blocks the transmission of light for some wavelengths and lets others pass. For instance, as seen through a red filter the world looks all red because only the light of the long 'red' wavelengths is transmitted (Figure 5.5; Plate 5.3). Similarly, the world looks blue through a blue filter, i.e. a filter transmitting only the short wavelengths. For many years, colour vision scientists have suggested that colour-blinds might use optical colour filters to compensate for some of their lack of biological colour discrimination.[170] This is, of course, directly analogous to the methods used for colour analysis in reproduction and photography. In a digital camera, the colour sensitivity of the three kinds of R, G and B pixels is set using optical filters with three kinds of transmission profiles (cf. Plate D.1). Which filters would be useful for a red-green blind person and how should he/she use them?

The problems experienced by red-green blind people are caused by deviating properties in one of the two kinds of longwave sensors: the L and the M cones have become too similar to each other or one of them is missing (Plate 5.2).

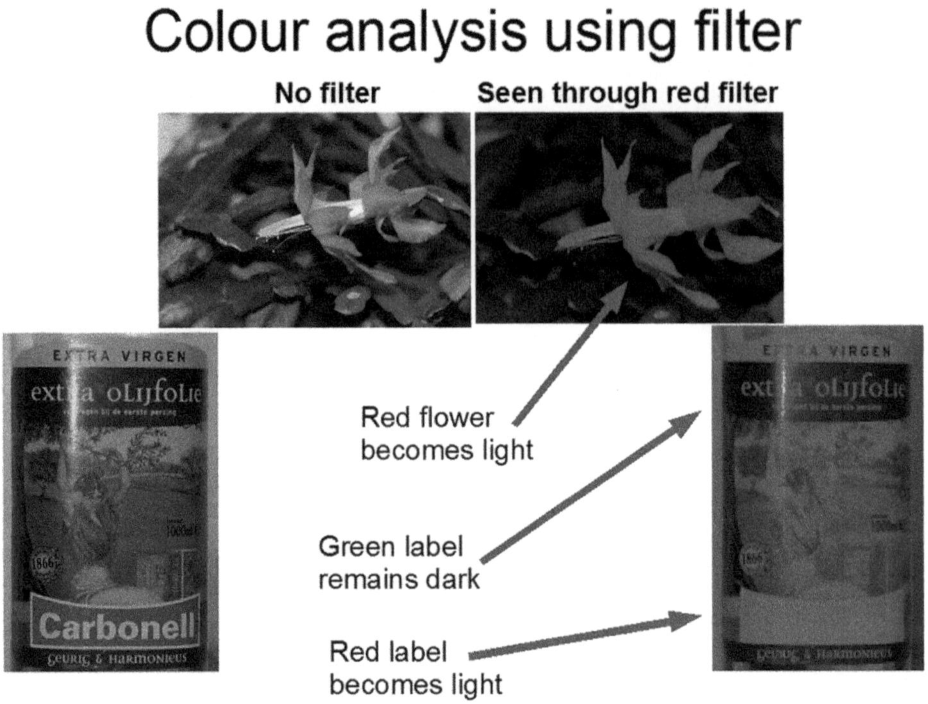

Colour analysis using filter

No filter   **Seen through red filter**

Red flower becomes light

Green label remains dark

Red label becomes light

Plate 5.3 Demonstration of how a red-green blind person might use an optical colour filter for analyzing differences between reddish and greenish colours. *On the left*: photographs taken without filtering, *Upper panel*: leaves and flower of cactus plant; *lower panel*: bottle label. *On the right*: the same scenes photographed with a red filter in front of the camera. The filter makes red items look relatively light and green items often become darker.
A black and white version of this figure will appear in some formats. For the colour version, please refer to the plate section.

Hence, colour discrimination becomes particularly difficult within the longwave region of the spectrum, e.g. between red–orange–yellow–green. This discrimination will not become better by permanently carrying an optical filter in front of the eyes. However, due to its selective wavelength transmission, a colour filter will cause a different change of lightness for different spectral colours. Such effects may be used as an aid to colour identification, comparing the relative colour lightness with and without the filter. For instance, your battery charger might emit a red light while charging and a green light when the battery is fully charged. To your naked colour-blind eyes, both LEDs might look equally yellowish. Then look at the light through a red filter: if it is still bright it is red, if it is drastically weakened or obliterated it is green. The same method can be used for the separation of reflected colours, e.g. as seen through a red filter an orange book will look pale and a green one will be dark (see Figure 5.5 and

Plate 5.3 for further information and examples). Thus, using optical colour filters one may, as it were, 'translate' some of the hue variations into differences of lightness, and our eyes and brains are very good at detecting such differences.

For a red-green blind person, it might be practical to carry a small coloured glass filter in the pocket. Suitable filters might, for instance, be acquired from shops for camera equipment or coloured (flash-) lights.[171] If you want to use the filter for colour analysis, then simply hold it in your hand and look at the coloured object with and without the filter. According to my own experience as a protanope (L blind), a single red filter is the most useful variety. I have also experimented with a two-filter combination: comparing how things look through a red filter, a green filter, and no filter;[172] however, I did not find that the green filter provided much useful extra information. Other kinds of filter have also been employed by various people, e.g. cyan or magenta (i.e. green or red filters that also transmit the short blue-violet wavelengths; Plate 2.8). The filter method for hue discrimination is coarse but practically useful, and it works best with strongly saturated colours, such as those often used for various kinds of colour-coded information (e.g. maps, tram-line diagrams, statistical charts, electrical leads, LED lights, electronic components,[173] etc.). It is less effective for analysing differences among the many subtle and often less saturated colours in nature.

A red-green blind person armed with a red filter might notice colour proper-ties that are less obvious to normal trichromats. For instance, and unexpectedly for myself, I found that blue wild flowers (e.g. blue bellflowers, Plate 2.13c) very commonly have a strong red component, which makes them seem light when viewed through a red filter; if they had been purely blue or violet-blue, they would instead have been dark through the filter. Another example concerns green leaves, which one might expect to become very dark as seen through a red filter. However, this darkening is often surprisingly slight and sometimes hardly noticeable. This unexpected reaction is due to the fact that, in addition to the 'green' intermediate wavelengths, green leaves also regularly have a strong red component in their reflected light and this fraction is not blocked by the red filter (cf. Plates 2.12c, 2.13b; Figure 5.5).

Instead of handheld filters, spectacles might be used which have had their lenses partly covered with a filtering stain (e.g. red in upper portions). One might then simply use eye/head movements for comparing how an object looks with and without a colour filter. Such kinds of eyeglasses have long been dis-cussed,[174] and they also seem to have been tested in New York in the 1920s. However, unfortunately, the method apparently never became sufficiently popular to make it available as a commercial item.

Besides optical filters intended for cameras, flashlights or other more general purposes, there are also commercially marketed filters specifically aimed at the colour blind. Such filters are often very expensive, and they are sometimes available as eyeglasses, sometimes as contact lenses for one or both eyes. When such a filtering contact lens is carried for only one of the eyes it could, of course, be used like the handheld filters described above: comparing the relative lightness of colours as seen with and without filtering by alternatingly looking through either eye.

As I have already noted, no permanent filter will improve the general colour discrimination of a colour blind. With a filter some colours will become lighter, others darker; some colour contrasts become stronger, others weaker; the number of distinguishable hues does not increase. Using a simple red filter, a colour-blind subject will often be able to read all the numbers in test plates like that of Plate 5.1; however, this does not at all mean that his colour blindness somehow was ameliorated by the filter. The filter simply changed some colour contrasts of the test plate into lightness contrasts, which may readily be seen by anyone, colour blind or not. Thus, when testing people's colour vision, it is extremely important to check that they do not have any coloured eyeglasses or contact lenses.

In special cases, red-green blind subjects might find a permanent colour filter to be useful, e.g. for enhancing the contrast between colours of importance for particular occupations. For instance, this may apparently be the case for sorting tasks involving fruit or tobacco leaves.[175] Besides, if somebody finds the world more agreeable as seen through a particular kind of colour filter, then this is a sufficiently good reason for using the filter.

Various kinds of optical filters might, of course, also be useful for other kinds of altered colour vision than the red-green blindness. However, the motivation to use such compensatory techniques would probably be less marked for the other main group of deviant colour vision whose hue confusions are mainly concerning blue/violet versus green/yellow (cf. Figure 5.3c). This category includes a small number of persons with inborn deviations and rather many people with acquired (temporary) changes of colour vision (Sections 6.1 and 6.3). In everyday life, people with such varieties of colour vision are generally less clearly handicapped than those with a red-green blindness; the essential colour codes of our society concern red versus green more often than blue versus yellow.

Besides the varieties of deviant colour vision, there are also groups of people who see no colour at all (Sections 6.2, 6.3.1). For such people the compensatory needs are much greater and more complex. However, also for this group of people, handheld colour filters might be of some use. The interesting autobiographical notes of the achromatic Norwegian physiologist Knut Nordby (1996), include a childhood memory of how he analysed differences between colours that he did not see, using a red-stained salad bowl as a colour filter.

5.7 Molecular biology and inherited red-green blindness: new perspectives

Most of our many inborn properties and functions depend on the collective effects of several different genes and, conversely, each gene tends to have an effect on several different functions. In such contexts, inborn red-green blindness is unusually simple: changes of single genes cause changes of well-identified proteins (L- or M-cone opsins, i.e. the protein part of the visual pigments) which are directly linked to easily determined single functions, i.e. the colour discrimination processes in retinal cones. Hence, inborn red-green blindness offers interesting opportunities for experiments concerning gene manipulations and gene therapy. In this context, two series of highly interesting experiments on animals have recently been published.

The first series of experiments concerned mice, which are normally dichromats with a colour vision rather similar to that of humans lacking the L cones (protanopes). Using standard techniques for genetic engineering, the investigators succeeded in providing mouse eggs with a human gene for the L-cone opsin. Using such eggs, a line of female mice were developed that had a different opsin gene for each X chromosome: one carried the gene for the normal mouse M opsin and the other one carried the gene for the human L opsin.[176] These female mice developed both M and L cones in their retina. Furthermore, remarkably, their nervous system managed to make use of the new L cones, giving the mice a more elaborate colour vision. For instance, the L+M-cone mice could react to the difference in colour between orange and green panels that looked precisely the same for normal mice. Thanks to their extra L cones, these engineered 'super-mice' had apparently obtained a trichromatic colour vision.

Another, perhaps even more astounding series of experiments concerned adult American squirrel monkeys (*Saimiri sciureus*). In each of their X chromosomes, such monkeys have only one opsin gene, i.e. one less than humans and Old World monkeys. Hence, all male squirrel monkeys are red-green blind dichromats because they have only one opsin gene in their only X chromosome (for further information, see Section 7.3). However, this single opsin gene has rather varying properties between various individual squirrel monkeys, corresponding to a human L or M or something in between. In monkeys that were missing the L-opsin gene, the colour vision was altered by introducing the gene for human L-cone opsin into their retinal receptor cells.[177] A virus was used for carrying the L-opsin genes, and a fluid containing the carrier virus was directly injected into both eyes, close to the retina and its receptor cells. Prior to this treatment, the monkeys had been trained to perform a test for colour vision, and they then showed normal squirrel-monkey properties, i.e. a colour vision corresponding to human dichromatic red-green blindness. At about 20 weeks after the gene injection the colour vision of the monkeys started to change, and they gradually developed a much improved and trichromatic colour vision.

The results of these latter studies indicate that, in the future, gene therapy might perhaps be used to treat human red-green blinds, possibly providing them with a more or less normal trichromatic colour vision. For a red-green blind dichromat, such a change would presumably be associated with the intriguing experience of perceiving completely new kinds of colours, such as 'red' and 'green'.

6 Other kinds of unconventional colour vision

Besides the commonly encountered inborn red-green blindness, there are several other categories of deviating colour vision, as will be described in this chapter.

6.1 Inherited blue-green blindness (tritanopia)

This is a relatively uncommon variety of colour vision with deviations concerning the 'third' kind of cones, the S cones (previously called 'blue' cones) (Table 6.1). People with an inborn blue-green blindness (*tritanopia*) may completely lack the function of these receptor cells. In some less extreme forms, the lack of S-cone function is evident but not complete. In addition, it has often been assumed that people might exist who have normal numbers of S cones although their colour vision is changed due to deviant S-cone functions, e.g. as caused by inborn changes in the properties of the S-cone visual pigment. However, it is uncertain whether such people and the corresponding colour vision category of *tritanomaly* actually exist.[178] The inheritance of inborn tritanopia is not sex linked; the gene for the S-cone opsin is localized to chromosome 7.

In CIE chromaticity charts, the confusion lines for tritanopes are radiating out from the lower-left blue/violet corner (Figure 5.3c); colours lying along one of these lines tend to be confused with each other. This kind of deviation has often been referred to as 'blue-yellow blindness'. However, as the confusion lines demonstrate, the uncertainties tend to concern blue versus green more often than blue/violet versus yellow; hence, this constitution should be called

Table 6.1 Occurrence of different types of deviating colour vision.

Type	%
Inherited deviations	
Red-green blindness	4.2
Blue-green blindness	~0.01
Rod achromatopsia	~0.002
S-(blue-)cone monochromacy	<0.001
Achromatopsia with normal acuity	<0.0001
Acquired deviations	probably > 5

Frequencies of occurrence given for Europe and for males and females together.
For references and further information, see Sharpe *et al.* (1999); Birch (2003).

'blue-green blindness'. Tritanopes confuse saturated colours of the wavelengths 400–510 nm, and they have a neutral point at yellow (~569 nm) and another one close to the violet shortwave end of the spectrum (<400 nm). People with this kind of deviation might have some use for an optical blue filter, as an aid for discovering differences between blues and shortwave greens. However, as was mentioned above, practical problems involving colour-coded signals are less severe for the blue-green blinds than for the red-green deficient members of the population.

6.2 Inherited total colour blindness

Total colour blindness (*achromatopsia*) is very uncommon; such people have no perception of chromatic colour and they see the world in terms of lightness variations only, in black and white and shades of grey, like in a black-and-white movie. Interestingly, even the concept of 'grey' might be difficult to grasp for an achromatic person. For us, 'grey' is mainly defined as something which has no chromatic colour, and chromatic colour is an unknown perception for the achromatic.

There are at least three categories of inborn total colour blindness (see Table 6.1 for statistics):

1. inherited rod achromatopsia;
2. inherited S-cone monochromacy;
3. inborn achromatopsia with normal acuity.

6.2.1 Rod achromatopsia

This is the largest of the various categories of total colour blindness. It is called *rod achromatopsia* (sometimes also *rod monochromacy*) because, in the most extreme cases, the vision of these people depends only on signals from the rods, i.e. the receptor cells that we are all using at night and under other low-light conditions. There are various transitional manifestations of achromatopsia in which the individuals still have small numbers of functioning cones. The more or less complete absence of functioning retinal cones leads, for the rod achromats, to several important consequences:

1. They see no colour, i.e. their vision corresponds to that of normal trichromats in deep darkness (*scotopic vision*), when using only the rods and not the cones.
2. Their light sensitivity varies with wavelength in the same way as it does for normal trichromats in the dark (see Figure 4.4, In darkness); i.e. red looks very dark and blue rather light. Due to this altered wavelength sensitivity, rod achromats may be able to read some of the numbers in common types of tests for colour blindness (cf. Plate 5.1); this happens in spite of the fact that they see no colours at all.
3. They cannot see well in strong light; in normal daylight they must protect their eyes with dark sunglasses in order not to become blinded. Similarly, rods of normal trichromats also do not function well in bright daylight; they become over-stimulated and are unable to react to differences in such high levels of lightness.
4. They have a low visual acuity, like we all have when seeing with our rods in the dark. This is mainly due to the manner in which rods are connected to the nervous system: many rods converge onto each nerve cell, which promotes a high light sensitivity and lowers resolution. For normal trichromats, the greatest acuity exists at the central fovea, a part of the retina lacking rods. In decimal scales of visual acuity, normal values are about 1.0, the minimal value required for a driving licence may be around 0.5, and rod achromats tend to show values around 0.1.
5. They have difficulties of visual fixation, and their eyes often show spontaneous movements (*nystagmus*) around the intended viewing direction. This is probably due to the lack of a functioning fovea, the normal centre of visual fixation.

Strangely enough, people with this extreme and very characteristic type of visual deviation often become wrongly diagnosed. Rod achromatopsia is quite uncommon, and the symptoms are often misinterpreted as being due to other causes for a decreased visual acuity.[179] In several countries, people with rod achromatopsia are well organized with their own societies and websites.[180] Under special conditions, in an isolated community with much intermarriage, the percentage of rod achromats might become relatively high. This is the case for the island Pingelap in the Pacific Ocean, and the life and conditions of these island inhabitants were described in a book by Oliver Sacks (1996). On his travels to Pingelap, Sacks was accompanied by the Norwegian physiologist Knut Nordby, who has himself published a detailed and interesting account of what it means to be a rod achromat (1996).

Genetically, different versions of rod achromatopsia exist, depending on changes within either the chromosomes number 2 or 8 (or, occasionally, number 14).[181] The inheritance is recessive, and the deviations occur about as commonly for men and women. For the achromatic inhabitants of Pingelap, the deviant gene belonged to chromosome number 8.

6.2.2 S-cone (blue-cone) monochromacy

As was described in Chapter 5, the function of either the L or the M cones are missing in inherited red-green dichromacy. Hence, one might assume that, sometimes, the inheritance might be such that the functions of both these types of cone are missing. Such a condition does indeed occur, but very rarely (Table 6.1). As the M- and L-opsin genes are located on the X chromosome, the inheritance of their functional absence is sex linked and recessive. S-cone monochromats are usually men who inherited their condition from their mothers. It is extremely uncommon among women.

At least two kinds of receptor cell with a different wavelength selectivity are required for colour vision. As the S-cone monochromats only have a single kind of cone they are completely colour blind. In addition, their visual acuity is very low, which is (at least partly) due to the low density and peculiar distribution of S cones within the retina. Only c. 5% of all cones are S cones (Section 4.3.1), and these cones are completely lacking from the most central part of the fovea, normally the site of maximal acuity at the centre of the visual field. For S-cone monochromats, the L and M cones are not only incapable of colour discrimination, but the low acuity indicates that they are altogether dysfunctional. In this respect, the S-cone monochromats differ from the red-green blind dichromats, the protanopes and deuteranopes. For these latter categories the visual acuity is normal, i.e. their lack of normal function for one type of the longwave cones is apparently not associated with a lowering of the general cone density, important for the detection of detailed lightness variations and contrasts. Thus, for the red-green blind dichromats, a normal total number of L + M cones seems to be available although, in these eyes, all are apparently filled with the same visual pigment (L or M).

6.2.3 Inborn total colour blindness with normal visual acuity

These individuals have normal visual functions with one striking exception: they see no colour, their perception of the world resembles a black-and-white movie. This deviation of colour vision is commonly referred to as *achromatopsia with normal acuity* or *atypical achromatopsia*. It is extremely uncommon (*c.* 1:1 000 000 or less) and, therefore, not well known with regard to its possible inheritance or underlying mechanisms. Furthermore, the nature of the deviation might well be different in different individual cases. For some scientists, a central localization seems most likely, and the condition is then called *cerebral achromatopsia*. The condition does indeed present symptoms resembling those that might appear in some uncommon and very localized kinds of brain damage (see next section). Thus, some of the inborn cases might depend on changes in cerebral organization. Alternatively, cases might conceivably exist with a peripheral mechanism, e.g. a combined inherited tritanopia and protanopia (or deuteranopia).

6.3 Acquired colour blindness

6.3.1 Changes of colour vision after brain damage

In the cerebral cortex of the brain hemispheres, the visual signals are first landing in the primary visual cortex, area V1. Subsequently, the processing continues in several other cortical regions with a more anterior position, including the V2 (secondary visual cortex) which lies along the anterior border of V1. A visual processing region of particular interest for colour vision is V4, lying still more anteriorly and on the underside of the hemisphere (Figure 4.8; note 123). Damage at V4 may cause a condition of total colour blindness (cf. Section 4.7). Such damage might be the result of external violence, or it might be caused by local cardiovascular problems (e.g. thrombosis). Such a loss of colour vision after brain damage is theoretically very interesting because it illustrates some essential aspects about how colour processing is done by the brain. After a brain lesion, the effects on colour vision might be localized to part of the visual field, or they might be very extensive, dominating the total visual experience. One famous case, with symptoms appearing after a car accident which probably caused localized brain damage, has been described by Sacks and Wasserman (1987). The patient, 'Mr I.', was an artist and he very vividly described how awful he found the world after it lost its colours. For instance, he had to eat with his eyes closed in order not to be nauseated by seeing the colourless food. One of the striking details concerns his purely mental responses involving colour. Prior to the accident, Mr I. had strong neuro-synesthetic experiences: he had always 'seen' patterns of vivid colours when listening to music. These mental colours also disappeared after the car accident.

At the level of single cells, the first steps in colour analysis happen in the retina: here the colour information of individual receptive fields starts to become organized into pairs of opponent colours (Plates 4.5, 4.6). The colour processing continues at various more central stations such as the LGN, V1 and more anterior regions (Sections 4.6.5, 4.7). In area V4 (Figure 4.8), many of the results of colour processing are apparently coming together and 'summarized', including complexities such as the calculations needed for colour constancy. This integrated 'colour package' may then be added, as it were, to the total visual perception of lightness patterns and contrasts, i.e. at a rather late stage in the processing it is joining the hard achromatic core of our visual capacities. Such an interpretation is strongly supported by the very fact that, in case of V4 lesions, all the colour components of the perceived image may apparently be removed without any disintegration of the image itself. Also without colours, visual images are perfectly recognizable and most of their objects and components can still be confidenty identified. Thus, in case of perceived visual images, colours are optional extras, nice and often useful things to have but not absolutely essential.

Interestingly, the loss of conscious colour vision after cortical damage may be combined with a retained ability to use colour contrast, even in the absence of luminance contrast, for discovering movements[182] and the outlines of objects' (edges).[183] Thus, also for colour vision there is an 'automatic' and subconscious variety which might, for instance, be used for the support of motor behaviour (cf. 'blindsight', Section 4.7.1).

Besides effects on the normal perception of colours in the surrounding world, variously localized brain damage may also give rise to other colour-associated symptoms. As mentioned above, synesthetic colours might disappear.

Conversely, all of the surrounding area might seem to be stained with a particular colour. This latter deviation is called *chromatopsia*, and specific names are also used for varieties involving different elementary colours (e.g. *erythopsia* for red; *xanthopsia* for yellow, etc.). Still other deviations are the inabilities to name colours (*colour anomia*) or to use colour terms (*colour aphasia*); such disabilities might prevail even though testing procedures show that the patient is capable of perceiving the differences between the various colours concerned.[184]

6.3.2 Changes of colour vision caused by disease, drugs, poison, ageing

Besides the changed colour vision due to brain damage, there are many other kinds of acquired colour vision defects. In general reviews and summaries concerning colour vision, it is sometimes stated that such effects are uncommon. However, systematic studies indicate the opposite; in the past such cases seemed uncommon mainly because they were not diagnosed. The acquired forms are often relatively mild, they often present few practical problems, and they are often not even noticed by the patients themselves. Thus, the occurrence of acquired changes of colour vision have to be systematically investigated, using sensitive and adequate tests (cf. Appendix A). The acquired deviations often (but not always) mainly concern the blue-green discrimination, resembling the inborn tritanopia (S-cone blindness). As mentioned above, blue-green blindness gives fewer problems in everyday life than the varieties of red-green blindness. Furthermore, in the past, many acquired cases of blue-green blindness might have been missed because some of the most popular colour vision tests are specifically aimed at discovering red-green blindness, leaving cases of blue-green blindness unnoticed (e.g. valid for the famous Ishihara test, Plate 5.1).

About 5% of the population is likely to have some kind of acquired deficiency of their colour vision.[185] Changes of this kind are common in ageing people, with or without a clearcut relation with various ageing-associated diseases. Furthermore, changes of colour vision occur in many kinds of eye disease, in various types of general disease, as a (side-)effect of medicines, drugs and poisons (Table 6.2). The emergence of a deficiency of colour vision might even be clinically useful as an early sign of complications in, for instance, diabetes or during certain kinds of pharmacological treatment.

Inborn deficiencies of colour vision are constant throughout life and, with extremely few exceptions, equal for both eyes. Acquired deficiencies are typically not constant but change with time, they are commonly different

Table 6.2 Examples of acquired deviations of colour vision.

Cause	Deviation characteristics
advanced age (and associated defects)	often lowered sensitivity for blue
lens increasingly opaque (cataract)	lowered sensitivity for blue
increased fluid pressure in eye (glaucoma)	discrimination problems blue/violet versus green/yellow
other eye diseases	often discrimination problems blue/violet versus green/yellow
diabetes mellitus with retinal damage (diabetes retinopathy)	discrimination problems blue/violet versus green/yellow
optic nerve lesions	often discrimination problems red versus green
brain lesions	varying characteristics and severity; in rare cases, complete loss of colour vision
side-effects of medicines, poisoning	varying characteristics and severity

This table lists only a few examples of the many observed kinds of acquired deviations of colour vision. Many of the deviations may show large variations in severity and duration, both for comparisons between different individuals and for symptoms of left versus right eyes in the same person.

between the two eyes, and they might become diminished or removed by a treatment of underlying causes.

In ageing people, the sensitivity for blue colours tends to decrease and there is an increasing tendency to confuse blue versus blue-green hues (tritan deviation). Such symptoms are partly due to an increased pigmentation of the yellow spot (*macula lutea*) at the centre of the retina, and partly to an emerging yellowish opacity of the lens. Such lens-changes are characteristic for cataracts, which belong to the major causes for acquired deficiencies of colour vision. The opaque yellowish lens functions as a yellow filter, blocking shortwave light, and this implies that blue colours will look progressively weaker as the cataract becomes progressively worse. Interestingly, in these cases colour vision may suddenly be restored to normal following a cataract operation: the removal of the yellowish lens leads to a sudden increase in the perceived colourfulness of blue. There are several documented cases illustrating how artists react to such alterations of their colour vision.[186] In some instances, the ageing painter uses progressively stronger shades of blue, apparently compensating for his/her decreased blue-perception and attempting to make the colours look as blue as they were remembered. The Swedish landscape painter Prince Eugen

(1865–1947) apparently belongs to this category. In other cases, the ageing artist might let the paintings become progressively less vividly blue as the cataract develops, perhaps exaggerating his experience that the blues of the surrounding world gradually became weaker than they were remembered to be. This seems to have been the case for Claude Monet (1840–1926). Furthermore, after his cataract was removed in 1923, Monet suddenly increased his use of blue colours in his paintings. These cases are intriguing because if an ageing painter with a developing cataract simply used the colours as he actually saw them while painting, similar strengths of blue should always have been present in the painting as in the motif; in both cases the artist would have seen the colours through the same yellow-filtering cataract lens.

Many kinds of poisoning may be associated with deviations of colour vision. Solvents for paint may apparently influence the colour vision of workers in paint factories. It has also been reported that colour vision deficiencies might occur after poisoning with, for example, carbon monoxide (CO) or lead. In the condition of *tobacco amblyopia*, a loss of visual acuity was typically combined with a deficient red-green discrimination. This condition was seen after long periods of intense use of tobacco (often in combination with alcohol); it does not occur much nowadays but it was formerly not uncommon among sailors at sea. Interestingly, the tobacco amblyopia was probably primarily due to vitamin issues (e.g. low levels of vitamin B_{12}) which might have caused retinal elements to become more sensitive to toxic effects of tobacco.[187]

Medicines which might lead to varying degrees of deficient colour vision (particularly at high doses and/or extended use) include:

- *digitalis*, a heart medicine, may cause deficient red-green discrimination, may also cause episodes of red or yellow chromatopsia (i.e. surrounding world drenched in colour);
- *monoamine oxidase inhibitors*, earlier often used against high blood pressure and endogenous depressions, may cause deficient red-green discrimination;
- *ethambutol*, used against tuberculosis, may cause deficient red-green discrimination;
- *chloroquine*, used against malaria and rheumatoid arthritis, may initially cause deficient blue-green discrimination, later on deficiencies in red-green discrimination and episodes of chromatopsia might occur as well;
- *Viagra*, used for erectile problems, may give a transient disturbance of blue-green discrimination and episodes of chromatopsia.

6.4 Variations in and above normal colour vision

Also among normal trichromats, differences occur in the characteristics of their colour vision and visual pigments. Thus, the opsin of the L cones exists in two normal versions with a known biochemical difference between them. Both versions are about as common, their functions are equivalent and their wavelengths for maximum light sensitivity differ by only a few nanometres.[188]

There has been much speculation as to whether some women might possess a 'super-human' colour vision, having four instead of three types of cone pigment. This might conceivably occur for female carriers of genes for deviant L or M cones (i.e. protanomaly or deuteranomaly). In their two X chromosomes, such women actually have the genes for three longwave opsins: the normal L and M plus one deviant L or M. In addition they have their normal S cones, i.e. a total of four kinds of cones with different visual pigments. Thus, paradoxically, the main candidates for excellent tetrachromatic colour vision are themselves carriers of genes for colour blindness.[189] However, research on these questions is still rather limited. As mentioned above, female carriers of genes for red-green colour blindness do indeed have slightly deviant colour vision;[190] however, most of these women may apparently still be classified as 'normal trichromats' and not as tetrachromats. Exceptions may exist: in a recently published study of female carriers for anomalous trichromacy, one of the 24 women did indeed display a seemingly tetrachromatic colour vision.[191]

These investigations suggest that, for most humans, the availability of four different kinds of cone pigment will not automatically lead to the establishment of a tetrachromatic colour vision. At first sight, such results would not be unexpected: the whole neuronal organization for the processing of colour vision would have to change in various complex ways in order to take four instead of three different colour sensors into account. However, surprisingly, such impressive feats of neuronal plasticity do actually occur in other species, as has recently been demonstrated in experiments with genetic engineering of opsins in mice and monkeys (see Section 5.7). Furthermore, in the normal embryological and childhood development of female American monkeys, the neuronal processing of colour vision automatically adapts to fit the number of available cone opsins, resulting in either a dichromatic or some kind of trichromatic colour vision (Section 7.3). It remains an intriguing question for forthcoming experimental investigations as to whether such a neuronal plasticity might also be present in adult red-green blind dichromatic humans, e.g. after genetic manipulations giving them three instead of two types of cone pigment (cf. Section 5.7).

A 'super-human' tetrachromatic colour vision, possibly present in some females, might seem a very attractive option, making it possible to see more colours than people normally do. However, such persons might, paradoxically, encounter communication problems rather similar to those of the colour-blinds, because the tetrachromatic colour-world and colour-nomenclature would not fit normal trichromatic standards. A tetrachromatic woman would presumably see a greater number of different hues than there are names for in our languages, and she would see differences between many colours that looked quite similar for most other people. A tetrachromatic designer might devise patterns of interior decoration that were visible and attractive only for other tetrachromats, but invisible and dull for normal trichromats (or for the dichromats).

7 Colour vision in different species of animals

For a better understanding our own visual capacities and our perception of colours, it is highly interesting to know what properties such functions have in other animals and how they emerged during evolution. Furthermore, for the general understanding of the behaviour and ecology of animals, it is in itself essential to consider the functions of their visual systems, including their colour vision.

7.1 Methods for comparative studies of (the capability for) colour vision

There are two main techniques available for studying colour vision capabilities in animals:

1. Behavioural experimental methods can be used for finding out how well an animal might distinguish different hues and levels of saturation. This is usually done such that the animal gets some attractive reward (piece of food or drink) if it makes a correct choice between a number of different colours. For instance, the food might be hidden behind one out of several panels, all painted in suitably different colours. If the animal pushes or taps on the correct panel, it receives the reward. When testing for the recognition of colour hues, it is important to take care that the animal cannot use any other clues, i.e. the hue must be recognized independently of the lightness and saturation of the target colour. Such behavioural methods are time consuming, partly because they typically require an extended period of training

before the animal learns what it is supposed to do. However, with patience and resolve, these techniques may deliver very detailed information about many aspects of the animal's sensory perceptions, e.g. which wavelengths and wavelength combinations may be distinguished from each other. Furthermore, with suitable adaptations, these methods are applicable to practically any species of animal; they have, for instance, been successfully used for studies of honey bees.

2. Physiological and anatomical/biochemical experimental methods can be used to investigate the functional properties of the sensory machinery available for colour vision. For instance, the reactions of the eye and its receptor cells may be measured while being stimulated with light of different wavelengths. With regard to colour vision it is then of particular interest to find out how many different types of receptor cells there are in the eye (e.g. kinds of cones and rods) and in roughly which relative proportions they occur.

   a. The most direct physiological method is to record the electrical reactions to various kinds of light stimulation in individual receptor cells. This is technically difficult, and such measurements are usually done *in vitro*, i.e. on eyes that have been surgically removed from the animal and kept in a bowl with a suitable salt solution. These experiments have often been done in animals such as fish and frogs.

   b. A coarser but still very useful technique is to record the electrical reactions of the whole eye (*electroretinogram*) while it is being stimulated with light of different wavelengths, intensities and durations. Such measurements can be done in intact animals, without any damage to the eye or its owner.

   c. For eyes from dead animals (or humans), microscopes may be used to analyse the absorption of light of different wavelengths in light-sensitive (unbleached) visual pigments of individual receptor cells (*microspectrophotometry*; cf. Plate 4.3b). Indirectly, such measurements provide information concerning the probable sensitivity of these cells for light of different wavelengths; only wavelengths that are absorbed by the visual pigment can have a physiological effect. Such microscopical investigations may also give a lot of interesting information concerning the relative numbers of the different types of receptor cell and their spatial distribution within the retina. These kinds of measurements have been very important for comparative studies of colour vision (capability) in different animal species.

Anatomical/biochemical and electrophysiological measurements (2c and b) are often more easily and quickly performed than the behavioural studies (1). If both categories of measurements are taken for the same animal species, the results are generally in good agreement, i.e. the colour vision of an animal may be well predicted by measurements of the properties of its eyes and retinal receptor cells.

7.2 Different animal species: evolutionary relationships and general properties of eyes and visual receptor cells

For the interpretation of comparative measurements in different animal species, it is important to know how the different species are related to each other. In the hierarchical subdivision of the animal kingdom, the highest category is called *phylum*. There are, in total, about 30 animal phyla in the world, and only roughly 10 of these have eyes capable of at least some (rudimentary) degree of image analysis, i.e. comparing intensities of light coming from different directions. Another third of the phyla has simpler organs with a sensitivity to light, and the remaining third has no light sensitivity at all.[192]

Our own species, *Homo sapiens sapiens*, belongs to the phylum of Chordates, subphylum Vertebrates, class Mammals and order Primates.

Only very few extant animals are chordates that do not belong to the vertebrates; this includes, for instance, the little swimming lancelets (*Amphioxus*). From an evolutionary point of view, the earliest vertebrates with still living representatives are jawless fish (e.g. lamprey) and cartilaginous fish (e.g. sharks, rays). Somewhat later during evolution, the ordinary bony fish appeared (*Pisces*), and from then on the evolution may be very coarsely described as occurring in the sequence: amphibians (e.g. frogs, salamanders) to reptiles (e.g. lizards, crocodiles, snakes) to mammals (including us). The birds (*Aves*) also developed from reptiles, and their evolutionary ancestors include the dinosaurs, which apparently sometimes also had feathers. All vertebrates have visual organs with the same general layout, camera-like single-chamber eyes with a nerve-cell-carrying retina and a single lens.

From an evolutionary point of view, the egg-laying mammals (echidna, platypus) and the marsupials (e.g. kangaroo, possum, etc.) are relatively ancient, retaining a number of primitive characteristics that are not present among other mammals. Our own order of primates have evolutionary ancestors in common with the tree shrews (*Tupaiidae*) and, at somewhat greater distances, with rodents (*Rodentia*, e.g. rats and mice) and lagomorphs (e.g. hares, rabbits, pikas).

Among the many invertebrate phyla of animals (Invertebrata), our closest evolutionary relatives include, strangely enough, the phylum of echinoderms (*Echinodermata*, e.g. sea stars, sea cucumbers). The extant species within this category often lack specialized organs for vision; sometimes their skin is pro-vided with a diffusely distributed light sensitivity which might be used for assisting behaviour. However, within several other invertebrate phyla different types of eyes have emerged and, independently of the vertebrate line, several different versions of visual systems have developed, including those with an advanced kind of colour vision (e.g. many insects and crustaceans).

Animals differ from each other with regard to the complexity of their system for colour analysis, e.g. the number of wavelength-selective receptor types used for analysing light composition (for vertebrates: number of types of cones). Examples of such differences are shown, for a variety of animals, in Plate 7.1. It should be noted that this graph only displays the wavelengths at which the various receptor cells and visual pigments have their maximal light sensitivity; for each class of receptor cell, the range of total light sensitivity covers a wide region at both sides of the peak (cf. Plate 4.3).

Animals also differ with regard to the total range of wavelengths used for vision. Many kinds of birds and insects, and also some mammals (e.g. rats), are capable of viewing the surroundings using ultraviolet light, at wavelengths shorter than those visible to the human eye. Our eyes are sensitive down to about 380–400 nm; animals with an ultraviolet sensitivity may see down to significantly shorter wavelengths (cf. Plate 7.1). Also, at the other end of the spectrum, differences in sensitivity exist between different animal species. Our own borderline between visible and infrared light lies at about 700–750 nm; some animals are also capable of seeing with light of significantly longer wavelengths (Plate 7.1).

7.3 Humans versus other primates, the evolution of trichromacy

Our closest animal relatives are the apes (e.g. gorilla, chimpanzee) and, at somewhat greater distances, the monkeys of the Old World (Africa, Asia). All these species of primates have a visual system and a colour vision very similar to that of humans: all are trichromats with three types of cones in their retina.

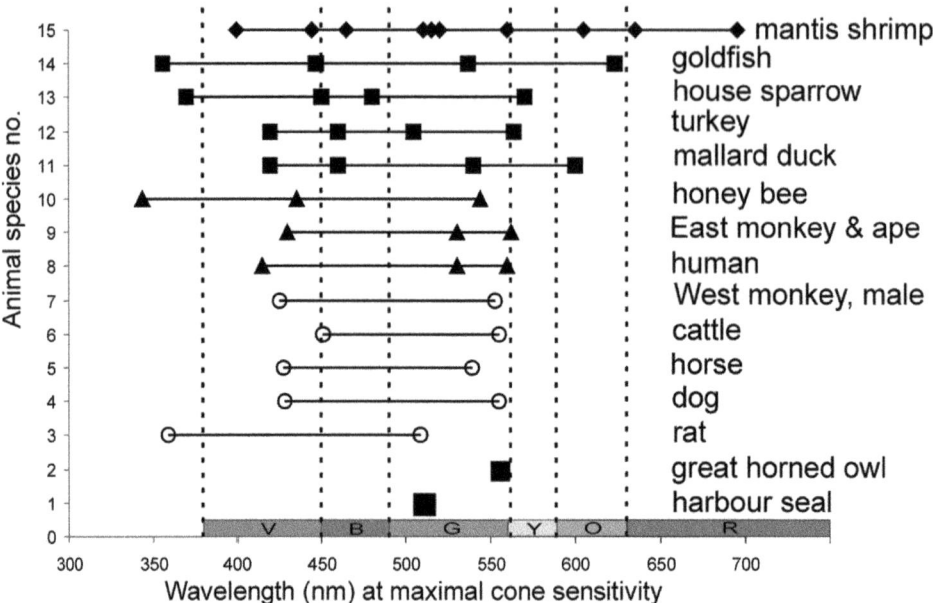

Plate 7.1 Comparisons between different animal species with regard to their conditions for colour vision. For each species, the wavelengths are indicated at which their visual cone pigments have their maximal light sensitivity and/or light absorption. Different symbols for different total numbers of cone types. For the two invertebrates (honey bee, mantis shrimp), the total number of different kinds of wavelength-selective light sensor is shown. The coloured bar along the x-axis shows the human range of light sensitivity, with vertical interrupted lines indicating the approximate transition-sites between different spectral colours. A black and white version of this figure will appear in some formats. For the colour version, please refer to the plate section.

Properties of visual pigments in different animal species:

1. Harbour seal (*Phoca vitulina*), Crognale *et al.* (1998)
2. Great horned owl (*Bubo virginianus*), Jacobs *et al.* (1987)
3. Rat (*Rattus norvegicus*), Jacobs *et al.* (2001)
4. Dog (*Canis familiaris*; greyhounds and poodle), Neitz *et al.* (1989)
5. Horse (*Equus caballus*), Carroll *et al.* (2001)
6. Cattle (*Bos taurus*), Jacobs *et al.* (1998)
7. Male American monkeys, in most species the male individuals are dichromats with varying properties of the pigment for longwave light; the illustration shows approximate means from data of Jacobs (1996)
8. Human, average values, see Section 4.3.1
9. African and Asian monkeys and apes, Jacobs (1996)
10. Honey bee, worker (*Apis mellifera*), Peitsch *et al.* (1992)
11. Mallard duck (*Anas platyrhyncos*), Varela *et al.* (1993)
12. Turkey (*Meleagris gallopavo*), Hart *et al.* (1999)
13. House sparrow (*Passer domesticus*), Chen and Goldsmith (1986)
14. Goldfish (*Carassius auratus*), Palacios *et al.* (1998)
15. Mantis shrimp, data for 10 visual pigments of the species *Gonnodactylaceus mutatus*, see Table 1 of Cronin *et al.* (2000).

Furthermore, all these primates have cones with very similar wavelength ranges for their light sensitivity and absorption. However, intriguingly, colour vision is organized in a different way for most of the American monkeys.

Only one species of American monkey, the howler monkey (*Alouatta*), is known to have a trichromatic colour vision similar to the Old World monkeys. As far as is known, all other American monkeys have a genetic constitution that causes the males to become red-green blind dichromats while the females show a wide range of variation, including cases of red-green blind dichromacy, cases of anomalous trichromacy and cases of normal trichromatic colour vision (i.e. a colour vision similar to that of humans and Old World monkeys). These deviant properties of American monkeys primarily depend on one basic genetic difference between them and the Old World monkeys: in Asia and Africa, monkeys have genes for each one of two types of longwave cones (L and M) in each one of their X chromosomes, while in American monkeys only one such gene is present in each X chromosome. Furthermore, American monkeys have a remarkable variability in their gene pool for longwave opsins: the single opsin gene on their X chromosome might have properties like the human L or M or something in between. Due to these circumstances, a male American monkey will always have the colour vision of a red-green blind dichromat (protanope or deuteranope or something in between), because he has only one X chromosome and one gene for a longwave cone. A female American monkey has, of course, two X chromosomes, and this gives her two genes for longwave cones. She may then have bad luck, getting two identical longwave genes and become a red-green blind dichromat. Or she might be lucky, getting the genes for a typical L cone plus that for a typical M cone, which would make her a trichromat with a colour vision like that of most humans.

It is thought that the situation for American monkeys represents an earlier evolutionary stage, ultimately leading to the uniform trichromacy of Old World monkeys, which requires a change from one to two longwave opsin genes per single X chromosome. It is known that, within a chromosome pair, genes might become displaced from one chromosome to the other. In a female living roughly 30 million years ago, the two single opsin genes of a pair of X chromosomes both seem to have landed on one of the X chromosomes. This 'super-chromosome', including both an L and an M opsin gene, was then passed on to the following generations. The resulting inherited capacity for trichromatic colour vision was of sufficient evolutionary value to become established as a normal property in later monkey generations. And, subsequently, in apes and humans as well. This process might, of course, have happened several times independently of each other; apparently it also happened at least once in America, for the howler monkeys.

Among humans, a surprisingly high percentage of red-green blindness still remains in the population; apparently this trait has not been sufficiently disadvantageous to cause its evolutionary removal. In this respect, humans seem unusual among the Old World primates. Inherited red-green blindness occurs also for monkeys in Africa and Asia, but very sparingly so. Using DNA analysis, evidence for red-green blindness was searched for among large numbers of macaque monkeys.[193] Out of 744 male and 1301 female crab-eating macaques, only three genetically colour-blind individuals were found, all of them male protanopes (L blind). Thus, apparently only about 0.4% of the male crab macaques were colour blind, a figure much below that for humans (Table 5.2). Among several other species of macaques, no DNA-signs of colour blindness were found (in a total of 455 males and 653 females investigated). In another study, similarly negative results were obtained in direct measurements of how monkeys reacted to different wavelengths of light.[194]

7.4 Mammals other than primates

As compared to other classes of vertebrates, the mammals have an unusually simple kind of colour vision: most mammals have only two kinds of retinal cones, one for shortwave and one for longwave light. Such is the situation, for example, for dogs, horses and cows (Plate 7.1). In popular media, it is sometimes stated that most mammals see no colours at all, which is totally wrong. The most common type of colour vision among mammals corresponds to that of a red-green blind dichromat (protanope or deuteranope). Thus, one should not try to train dogs and horses using red–orange–green signals (blue versus yellow should be fine). The red cape used by bull fighters to excite the bull will probably be seen as coloured also by the bull, but its colour would be less unique for the bull than for the matador, and the animal will probably be more irritated by the movements of cape and matador than by the colour of the piece of textile.

Only a very few species of mammals, with a highly specialized kind of life, have eyes that probably lack colour vision altogether. This includes animals with only one type of cone in their retina, e.g. toothed whales and seals (Plate 7.1).[195] In such instances there is still a remote possibility that colours might be seen under mesopic conditions, using rods and cones together; however, no such case has yet been proven to exist. Apparently, no species of mammal has a complete lack of cones, i.e. there are no mammalian 'rod achromats'. For instance, moles, living in almost perpetual darkness, have a visual system with very few receptor cells, but they still have two types of cones, which should enable them to see colours and, most importantly, to see anything at all when they occasionally come to the surface in normal daylight.[196] However, even if most mammals have the sensory equipment needed for colour vision, it is not self-evident that they all make much use of it. For instance, cats are dichromats and capable of discriminating colours, but in their daily behaviour they do not seem to care very much about these phenomena.[197]

Mammals with a trichromatic colour vision belong to one of two very different groups of species:

1. Some marsupials, i.e. evolutionarily ancient kinds of mammals.[198] However, the even more ancient egg-laying Australian platypus is a dichromat.[199]
2. Various primates, i.e. humans, apes, monkeys of the Old World, American howler monkeys (see above, Section 7.3).

With regard to the evolutionary ancestors of monkeys, the lemurs of Madagascar are particularly interesting. For one species of lemur (*Lemur catta*), behavioural observations suggested a trichromatic colour vision. However, physiological measurements showed that these lemurs were actually dichromats (two kinds of cones) but, under appropriate light conditions (mesopic vision), they were capable of using cones and rods together for their colour vision.[200]

7.5 The vertebrate ancestors of mammals: reptiles, amphibians and fish

Mammals developed from reptiles, which came from amphibians, which came from fish. In everyday language, words like 'development' are generally felt to imply and include an element of 'improvement'. Hence, one might intuitively expect that the evolutionary ancestors of the mammals had only a rather primitive kind of colour vision, as dichromats or worse. However, this is not the case; on the contrary, many of the present-day species of reptiles and fish have an excellent colour vision, exceeding that of humans and other primates. Reptiles and fish often have eyes provided with four or more types of cones, each one with a different visual pigment. Among the most primitive kinds of present-day vertebrates, the jawless fish (*Agnatha*), some species have five different visual cone pigments,[201] suggesting a pentachromatic colour vision.

(a)

More

(c)

Similar

(e)

Less

(b)

(d)

(f)

Plate 7.2 Complexity of colour vision in different species, as compared to that of humans. (a-b) Mallard duck and goldfish are tetrachromats. (c) Rhesus macaque; monkeys and apes from Africa and Asia have a trichromatic colour vision very similar to that of humans. (d) The honey bee is also a trichromat, but it is capable of seeing much further into the ultraviolet than is possible for humans (cf. Plates 7.1 and 7.3). (e) American brown spider monkey; for most American species of monkeys, males are dichromats and females have varying capacities of colour vision (section 7.3). (f) Cows and most other mammalian species are dichromats. (*Photographs included from Wikipedia by: (c): J.M.Garg; (d): Bartosz Kosiorek Gang65; (e): http://www.birdphotos .com edit by Fir0002; for further source information, see p.xiv).*

The chameleon is a reptile with particular colour properties: its skin contains different coloured elements which quickly may become enlarged or diminished, causing the animal to alter its colour. Such changes might have social functions in interactions with other chameleons, and/or they might serve as camouflage. As one might expect, chameleons have a well-developed colour vision, supported by four different cone pigments.

Among fish, the complexity of colour vision is related to their way of life and normal surroundings. An advanced tetrachromatic colour vision is, for instance, present among tropical fish living close to the surface. Such fish are themselves also often brightly coloured (e.g. goldfish, Plates 7.1, 7.2). On the other hand, fish of muddy freshwaters are often dichromats, and deep-sea fish, living in darkness, often have only rods and see no colour at all.

Certain species of snake have in their head a special *pit organ* that is sensitive to heat radiation, i.e. rays with wavelengths of about 5000–30 000 nm, far longer than those of visible red light (about 635–750 nm). However, this pit organ is not a regular image-forming eye, but rather a primitive visual device for discovering the presence and rough direction of heat radiation from an interesting warm-blooded prey, e.g. a mouse or bird.

In the course of evolution, the dichromatic colour vision of mammals apparently emerged due to simplifications of a more advanced earlier system, perhaps in association with the nightly habits of many mammals; good rod vision and an excellent sense of smell were perhaps more important than a richly developed colour vision. Furthermore, the poor colour vision of most mammals is in apparent agreement with their own exterior, typically rather colourless as compared to birds and tropical fish.

7.6 Birds

Birds also developed from reptiles, i.e. from an evolutionary point of view, birds are cousins of mammals. Their reptilian ancestors probably had good colour vision, and this has been well preserved in birds. In flying birds, vision is of extreme importance for motor and behavioural control, and colour vision is likely to be of importance as well, e.g. for orientation and recognition of landmarks and for the identification of appropriate food.

Daylight birds often have four types of cones and a tetrachromatic colour vision, i.e. better than most people. Birds are often capable of seeing further into the shorter wavelengths (i.e. into ultraviolet, UV) than we are. In the eyes of birds and reptiles, the wavelength selectivity of the various cones is based not only on the absorption properties of their visual pigments, but also on the effects of coloured oil droplets that are present in the light path of the cones.[202] Such a droplet works as an optical filter, narrowing the range of wavelengths used for activating the visual pigment of the same cone. In such cones, the wavelength at which the cell has its greatest sensitivity depends partly on the properties of the droplet filter and partly on the absorption characteristics of the visual pigment.

Also among birds, differences in colour vision are often associated with differences in the way of life. Night birds have less need for colour vision, and some kinds of owl seem to have eyes with only one type of cone; such birds are probably completely colour blind (e.g. the American great horned owl,

Plate 7.1[203]). In other species of owl, small numbers of other kinds of cone have been observed, but it is still unclear whether these cones might suffice to give the owls a useful degree of colour vision. All kinds of bird seem to possess at least one type of cone.

7.7 Invertebrates

Among the invertebrates, image-forming *single-chamber* eyes exist in, for instance, squids, octopuses and spiders. However, the evolution of these eyes happened completely independently of the evolution of vertebrate eyes. In the evolution of single-chamber eyes, the most primitive initial stages concern eyes in the shape of a pit or vesicle with a floor covered with light-sensitive cells. One or several such eyes might give an animal a useful but very simplified dynamic 'picture' of its surroundings: showing how the light is distributed around the animal, in which direction it is changing, where it becomes stronger or weaker. Insects and crustaceans often have another kind of eye, *compound eyes*, which are composed of a large number of separate 'eye units' (*ommatidia*) placed beside each other, each one with its own lens and receptor cells. Also among the invertebrates, a sensory machinery appropriate for colour vision is often present, i.e. several types of receptor cell with differing wavelength selectivities in their light sensitivity and absorption.

7.7.1 Invertebrates with camera-like eyes

The best-known invertebrates with single-chamber eyes are the cephalopods, i.e. squid and octopus. Like ourselves, such an animal has two large spherical eyes with a lens, an iris and a retina containing light-sensitive cells. From some points of view, the cephalopod eye seems more logically constructed than vertebrate eyes. In our eyes, the light-sensitive parts of the receptor cells are lying toward the eye's outer wall, and the light typically has to pass through all the layers of nerve cells and through the cell bodies of the receptor cells before finally reaching the visual pigment (Figure 4.2). In the cephalopod eye, the receptor cells are more 'correctly' placed with their light-sensitive portions directed toward the incoming light, and the nerve fibres are all lying behind the receptor cells, not obstructing the light; there is no 'blind spot' in a cephalopod eye. The processing of the visual signals occurs in assemblies of nerve cells (*ganglia*) lying outside the eye. Cephalopods have an excellent vision with regard to lightness contrast. They are, however, paradoxical creatures with regard to colour vision. Like chameleons, cephalopods are capable of rapid and dramatic changes in the colour and lightness patterns of their skin, used and useful for a very effective camouflage as well as for social interactions. In spite of these abilities, most species of cephalopod have only one type of visual pigment in their eyes and are completely colour blind.[204] Their own visual monitoring of their camouflaging skin changes is apparently limited to the (re-)viewing of lightness variations.

Camera-like eyes occur also in certain other species of invertebrates; interesting examples are spiders, which generally have eight eyes with differences in function and viewing direction. In particular, the hunting spiders (e.g. jumping spiders) have good vision and, apparently, also good colour vision. In their retina, sensory cells with different kinds of visual pigment lie at different distances from the lens, organized such as to diminish the problems of chromatic aberration.[205]

7.7.2 Invertebrates with compound eyes

The large phylum *Arthropoda* includes animals with an exoskeleton, e.g. insects, spiders, crustaceans. Adult insects and crustaceans usually have compound eyes, consisting of a large number of separate *ommatidia* (see Section 7.7). There are two main types of compound eyes:

1. *apposition eye*, the common one for daytime insects and many crustaceans;
2. *superposition eye*, common among nighttime insects and for shrimps and crayfish living at depths with weak light.

In the apposition eye, each ommatidium works independently. Each ommatidium has several receptor cells (usually eight) which are bundled together into a *rhabdom*, within which the various receptor cells all have the same visual field but (partly) different wavelength sensitivities. In such an invertebrate eye, the colour sensitivity is present within each pixel and not, as in our own eyes, distributed across several pixels with different locations. Insects often have good colour vision, sometimes available also in very weak light. Flowers use colours as signals for attracting pollinators, and a good colour vision is then, of course, very useful for pollinating and nectar-collecting insects like bees and butterflies. Most insects are trichromatic

Plate 7.3 The same flower may show different brightness patterns in light visible for our eyes (left), and in ultraviolet light that is visible for the honey bee (right).
Source: http://en.wikipedia.org/wiki/File:Mimulus_nectar_guide_UV_VIS.jpg.
Image by Plantsurfer.
A black and white version of this figure will appear in some formats. For the colour version, please refer to the plate section.

(including the honey bee, Plates 7.1 and 7.2d),[206] and almost all insects have a sensitivity extending into the UV region of short wavelengths. Such a UV vision is useful for recognizing flowers with a UV-patterned kind of design (Plate 7.3). The honey bee and many other insects are rather insensitive to the long wavelengths of reddish colours (>600 nm), but there is much variation and insects exist with a sensitivity for an extremely wide range of wavelengths (e.g. <300 to >700 nm).

Some butterflies and dragonflies have tetra-, penta- or sextachromatic colour vision, while crayfish and crabs often are tri- or dichromatic.[207] The world record with regard to the complexity of colour vision is currently held by a strange hunting crustacean, the mantis shrimp (*Odonotodactylus*; Plate 7.4). It is distantly related to ordinary shrimps, but it is larger and hunts by hiding and waiting for its prey. The mantis shrimp is apparently remarkably similar to the *Anomalocaris* (= deviant shrimp), a carnivorous animal living in the Cambrian seas about 525 million years ago. Across each one of its large apposition eyes, the mantis shrimp has rows of ommatidia, each one with different types of receptor cell at different depths. For these complex eyes, the mantis shrimp uses many different visual pigments with differing wavelength sensitivity. For the species in Plate 7.1 (*Gonnodactylaceus mutatus*) the number of pigments shown here is only 10, but in other species of mantis shrimps as many as 16 or more have been observed.[208] The animal analyses the visual field and its colours using a series of strange eye movements, letting the narrow stripes of different colour-sensitive cells be reached by light from all

(a)   (b)

Plate 7.4 An international champion of colour vision complexity: the *mantis shrimp* (*Odonotodactylus*). (a) Body anatomy, as seen from above. (b) As seen alive in natural surroundings. See Plate 7.1 for an example of its number of receptor-types with different wavelength sensitivities.
Sources: (a) http://en.wikipedia.org/wiki/File:MantisShrimpLyd.jpg; (b) http://en.wikipedia.org/wiki/File:Mantis_shrimp_from_front.jpg, image by Jenny.
A black and white version of this figure will appear in some formats. For the colour version, please refer to the plate section.

the interesting details of the scenery. Recent behavioural investigations have shown that this 'multichromatic' animal indeed reacts differently to different colours but, paradoxically, its powers of spectral wavelength discrimination are rather poor, much coarser than the discrimination capability of trichromatic humans.[209] The brain and visual system of the mantis shrimp are apparently not processing its complex visual information for achieving our kind of gradual and nuanced colour vision, but perhaps rather for the purpose of rapid colour recognition. Thus, for instance, the animal might be using each one of its many colour receptors for the rapid identification of an observed colour as to whether it belongs a particular (and behaviourally interesting) 'colour group'.

7.8 Sensitivity to light polarization

Besides differences in wavelength composition, light rays might also differ in the directions of their wave movements. Normally, light waves are oscillating toward all sides around the line of the ray's general direction, i.e. left/right, up/down, etc. When a light beam is obliquely reflected against the non-metallic surface of an object (e.g. glass, water), the oscillations will diminish in some directions but not in others. In such a case, the reflected light has become *polarized*. Sunlight also becomes polarized when the rays are colliding with air molecules, scattering shortwave light such that the sky becomes blue.[210] As seen from the ground, the sunlight polarization is rather weak close to the sun and it gradually increases with distance until a maximum is reached for regions of the sky at about 90° from the sun.

The degree and direction of light polarization may be observed using a *polarization filter*, a material which only transmits light with one direction of polarization. If a polarized obliquely reflected light is viewed through such a filter, its intensity will become drastically changed if the filter is rotated by 90°: in one position all the obliquely reflected light (e.g. reflections in a glass window) seem to be transmitted, in the other filter position the oblique reflexes may be almost gone. The light from most of the scenery is not polarized and will be equally well transmitted with the filter in either position. Such polarization filters are often used in photography, e.g. useful when picturing something through a window without being bothered by reflexes in the glass panes. For similar reasons, polarization filters are also sometimes used in sunglasses (e.g. helpful against reflections of sunshine on water along the beach).

Visual pigments are sensitive to the relation between the polarization-direction of light (if any) and the orientation of the pigment molecule: if the light is polarized, the pigment only becomes activated if the light oscillations occur in the correct direction for the molecule in question. In vertebrate eyes, like our own, the pigment molecules of the receptor cells are oriented in all possible directions. Hence, we cannot see any differences in light polarization; a polarized light looks the same intensity irrespective of how we rotate our head and eyes. However, for some invertebrates the situation is different. For instance, in insects and crustaceans with compound eyes, the visual pigments might be geometrically ordered to a degree that makes the whole receptor cell sensitive to the direction of light polarization.[211] Some insects apparently use their polarization sensitivity for recognizing reflecting water surfaces. Behavioural studies of honey bees and ants have shown that, looking at the sky, they are capable of using their polarization sensitivity for calculating the position of the sun even if it is covered by clouds.

Appendix A   Diagnosis and measurement of differences in colour vision

Methods for the testing of colour vision became needed when it was discovered, at the end of the eighteenth century, that a considerable percentage of the population was 'colour blind'. In the earliest investigations, the testing material often consisted of various colours of cloth. One such test, 20 ribbons in different colours, was used by John Dalton, who asked his participants about the colours of these items in daylight as compared to in candle light (cf. Section 5.1). In the 1870s, Frithiof Holmgren developed a test with skeins of coloured wool (see note 168), using procedures based on colour matching rather than naming. A large number of variously coloured skeins were mixed in a heap. The examiner chose a small number of 'master skeins' and then asked the participant to search the heap for other skeins with a similar colour.

Nowadays deficient colour vision is commonly tested with several techniques used in parallel.[212] For a rapid and easy initial diagnosis, one of the many 'coloured-dot tests' may be used (Section A.1). For persons found to be colour blind in such a test, it will often be appropriate to add some other procedure for evaluating the degree of colour vision deviation. This might involve a general colour-sorting task, like the D15 test (Section A.2). For certain kinds of occupations, potential employees will be evaluated using colour-identification tasks similar to those encountered in a work situation (e.g. coloured light signals like those used at sea; Section A.3). In scientific contexts, it is important to be able to identify both the degree and the type of deviation; this is best done using a specialized instrument called an anomaloscope (Section A.4). Still other procedures might be added in the future (see Sections A.5 and A.6).

A.1 Pseudoisochromatic plates

A.1.1 Ishihara

One of the best and most famous examples of this type of test was developed in Japan by Shinobu Ishihara (1879–1963), a professor of eye diseases at the

University of Tokyo. The first edition of his test was published in 1917, around 100 years ago. The test is so well made that it is continuously reprinted, and it remains one of the most frequently used tests for colour blindness. The test plates consist of variously coloured dots, organized such that those of a particular colour form a pattern in the shape of numbers (Plate 5.1). The colours of these patterns and those of the surrounding 'background' dots have usually been chosen such that they look different for normal participants but resemble each other for the colour-blinds (cf. the confusion lines in Figure 5.3a, b). Furthermore, the relative brightness of the dots varies randomly such that the hidden digits cannot be seen as patterns of luminosity contrast (*random luminance masking*). Thus, for colour-blind participants the dot patterns and their backgrounds are indistinguishable, and no numbers can be seen, because the dots all seem 'isochromatic', i.e. having the same colour. However, the dots are not truly isochromatic because for the normal trichromat they have clearly different colours; hence, the dots of such test plates are said to be 'pseudoisochromatic'.

The Ishihara test is very sensitive and it may diagnose many participants as colour blind who have few or no colour vision problems in daily life. It is important to remember that this test was made for the specific purpose of identifying people with red-green blindness. For the diagnosis of other types of colour blindness, different pseudoisochromatic test plates are required (e.g. for cases of inherited tritanopia and similar kinds of acquired deficiencies of blue-green discrimination; see Section A.1.2).

Besides the test plates in which the patterns are invisible for the red-green blind ('vanishing design', Plate 5.1), the Ishihara test contains a control plate in which the digits can be seen by all participants: this plate demonstrates how dot patterns are organized to form digits, and it may also be used for the identification of malingerers. Furthermore, there are several plates in which different digits are seen by normals and by colour-blinds ('transformation design'). During a testing procedure, the existence of such plates might seem encouraging for the colour-blind participant, finally some numbers are visible! However, unfortunately, such 'colour-blind' digits are not visible thanks to some unexpected visual capacities: in plates of transformation design, the digit patterns may simply be formed by dots of different confusion colours used together, all looking the same for the colour deficient person but not for normal trichromats. In some Ishihara plates, only the 'colour-blind' digits are shown and nothing is seen by the normals ('hidden design'). In still other test plates, the dots are coloured and arranged such that people with L- and M-cone abnormalities see different digits ('classification design').

Each Ishihara plate should be viewed at a distance of about 75 cm for c. 4 s or less. The complete test contains a total of 38 plates, but it is often used in abbreviated versions with 24 or 14 plates. In all tests of colour vision, whatever the procedure, the test leader must make sure that the participant does not carry coloured eyeglasses or contact lenses (see Section 5.6.3). Furthermore, it is important to perform the tests using a 'neutral' illumination similar to daylight. For instance, test plates illuminated by a reddish light will look as though viewed through a red filter; the plates will no longer be 'pseudoisochromatic' for the red-green blind and such participants might then possibly read the test digits.

A.1.2 Other pseudoisochromatic tests

Ishihara was not the first to design this type of test for colour vision. His main predecessor was the German ophthalmologist Jakob Stilling (1842–1915), who in 1877 launched a similar test called the 'Stilling Colour Chart' or 'Stilling's Colour Table'. Apparently it took some time before his test became widely accepted; people found it unbelievable that a proper test would diagnose so many 'intelligent and cultivated persons' as being colour blind.[213]

There are many other pseudoisochromatic tests, either useful as replacements for the Ishihara test, or valuable for supplementing it in various ways. A general test that has been widely used was developed by doctors Hardy, Rand and Rittler for the American Optical Company (first edition 1954).[214] This HRR-test contains 24 plates with dots in variously saturated colours forming patterns of geometric shapes (circle, triangle, etc.) against a background of grey dots. Besides separating deficient from normal colour vision, this test also provides suggestions as to the approximate degree of the disturbance and whether it concerns L, M or S cones. Another widely used pseudoisochromatic test is the British 'City University Tritan Test' (5 test plates), designed for the diagnosis of blue-green vision deficiencies and often used as a supplement beside the Ishihara test.

There are very many other tests of a similar general type,[215] and they are also richly represented at various sites on the internet. Via such sites, it is easy to get a general impression of how pseudoisochromatic tests are constructed. Often the internet surfer may test his own eyes, which is useful as a general experience and, perhaps, provides suggestive evidence concerning the state of his own colour vision. However, it should be remembered that colour vision tests performed on one's own computer would provide reliable results only if the test had been preceded by a very careful calibration of the colour rendition of the monitor screen. For similar reasons, test plates reproduced in general non-scientific books and magazines should be used with caution; the exact repro-duction of colour nuances is still technically difficult (cf. Section 2.6).

A.2 Arrangement tests: Farnsworth D15 and 100 hue

This general kind of testing method has been particularly important in employment contexts. One of the best known and still often used examples is the *D15 test*, which was developed in 1943 by Dean Farnsworth, initially as an aid for the selection and employment of electricians at a transformer firm. In this test, the participant is given 16 differently coloured 13-mm buttons, one of which is defined as the starting point and the other 15 as test buttons, all of them numbered on the reverse. The participant must begin by searching for the second button, i.e. the one having a colour most closely resembling the starting button. After that he should select the third button, being the one most closely resembling the second button, and so on, until all the 15 test buttons have been arranged in the best possible colour sequence (cf. Plate A.1). For the normal trichromat, all the buttons have colours of the same lightness and saturation, and their hues vary in a manner generally corresponding to the colour circle of the Munsell system (Section B.1.1; Plate B.1). The D15 test is devised such that persons with only a slight deviation of their colour vision may pass the test with no errors, i.e. arranging the buttons like a normal trichromat. Even for normal trichromats, one minor error per session is accepted. Only about 5% of all Western European males will fail in the D15 test, although 8% of the males will fail in the Ishihara test. Thus, the combined D15 and Ishihara test results give a rough, but practically useful, gradation of the degree of deviation in colour vision. Furthermore, the D15 results will suggest whether the deviation concerns the L, M or S cones. People with a strongly deficient colour vision (e.g. dichromats), will arrange the D15 buttons in a markedly non-normal sequence with many errors (Plate A.1b, e), and they will also often arrange the buttons in a somewhat different sequence in repeated testing sessions. This behaviour also differs from that of normal trichromats, who will typically produce the same 'correct' D15 sequence in session after session.

   In arrangement tests, a practical concern is to keep the test buttons clean and their colours constant. Such problems have been minimized for test versions in

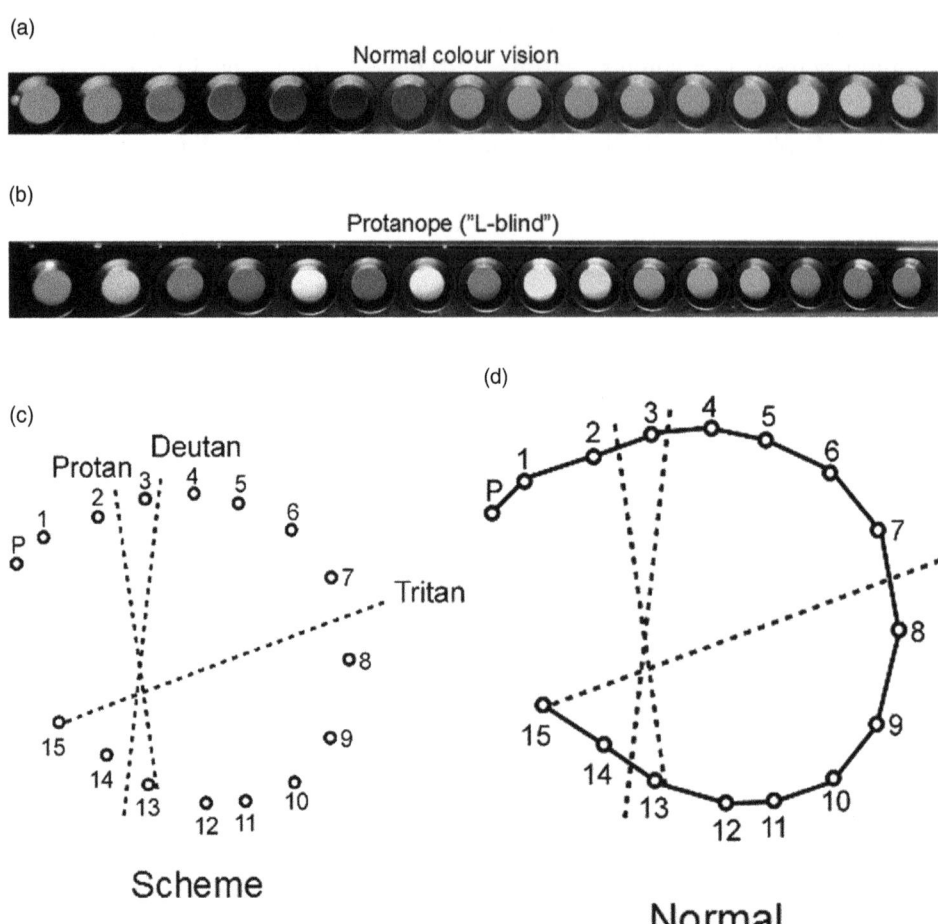

Plate A.1 The Farnsworth D15 test. **(a, b)** Buttons arranged in a colour sequence conceived as correct by a normal trichromat **(a)** and by a red-green blind dichromat (protanope; **b**). **(c)** Scheme for the evaluation of D15 results. As part of the analysis, lines are drawn between the numbered buttons according to the sequence in which they were arranged. **(d)** Evaluation scheme completed for test in (a) (normal trichromat): no 'crossing' lines. **(e)** Evaluation scheme completed for test in (b) (protanope): many 'crossing' lines. In the evaluation scheme, interrupted lines show the approximate slanting directions for 'crossing' lines, as seen in different types of deviating colour vision. A black and white version of this figure will appear in some formats. For the colour version, please refer to the plate section.
For further information, see p. 65 in Birch (2003).

which the buttons are never touched; they lie inside a translucent plastic container and may be moved around using a magnet.[216]

In the much more extensive *Farnsworth-Munsell 100 hue test* there were originally indeed 100 buttons for sequential arrangement, using principles like

(e)

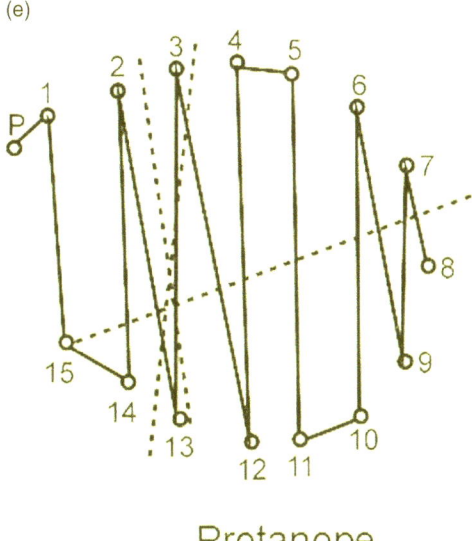

Protanope

Plate A.1 (*cont.*)

those of the D15 test. Later on, the number of test buttons was decreased to 85, but the test is still usually referred to as the *100 hue test* or *FM100*. This huge arrangement test is time consuming and mainly used in research contexts, e.g. for evaluating the precision of colour vision in normal trichromats. Even people with normal colour vision often make errors in the FM100; the older the participant, the more errors are made.

A.3 Practical tests: lanterns and wires

In many vocations, normal colour vision is needed because the occupation requires that red and green light signals can be reliably distinguished, e.g. in railway traffic, ship navigation and aircraft flying. For the screening of possible employees in such occupations, tests have been devised that imitate, as well as possible, the practical colour discrimination tasks involved. Light signals are then produced using carefully specified testing lanterns, and participants are asked to identify their colours when seen at different distances and at different light intensities. Such specialized testing lanterns have long been used; one of the earliest specimens was devised by Donders in 1877.[217] Different testing lanterns are used in different countries; in Britain lanterns of Giles–Archer or Holmes–Wright types are common, and in the United States those of type Farnsworth.

In some occupations, important technical components have to be recognized using a colour coding system. This has, for instance, long been the case for electrical cables and components. When screening people for the possible employment as electricians, practical tests have been used that resemble actual work tasks, e.g. interconnecting pairs of similarly colour-coded wires in two complex multistranded cables. For people with a light degree of colour vision deviation such a task might be easily performed; for a dichromat it might be impossible without using some kind of technical aid (e.g. an optical colour filter, Section 5.6.3).

A.4 Anomaloscope

This is an instrument for performing psychophysical measurements of a kind that may be used for evaluating both the type and degree of colour vision deviation. The instrument may look like some kind of old-fashioned telescope or, in more modern versions, somewhat resemble a large microscope. Looking into the eyepiece, the participant will see a circle split in halves (Plate A.2). One half is 589 nm yellow and its intensity may be adjusted. The other half shows a colour produced by the additive mixture of 670 nm red and 546 nm green light[218] and the proportions of red and green may be adjusted. For a particular mixture of red and green the result will look yellow; such a match between yellow and (red + green) is called a 'Rayleigh match' after the physicist Lord Rayleigh (Figure 5.1b). As the red-green mixture of the anomaloscope is adjusted from one extreme to the other, a normal trichromat will see a gradual colour change from red via orange and yellow to green. The participant is asked to adjust the (red+green) mixture such that the colour of this half-circle looks identical to the purely yellow one. If needed, the intensity of the purely yellow half may be adjusted.

For normal trichromats, (red + green) mixtures produce an acceptable yellow only within quite narrow limits, and similar mixture proportions are required for different individuals and for the same person at different times. Anomalous trichromats will usually accept a greater variation of (red + green) mixtures as yellow, and the range of this variation gives a measure of their colour vision deviation. In addition, anomalous trichromats will typically not perceive a matching yellow colour for the (red + green) mixtures accepted by normal trichromats: protanomals (L weak) require more red and deuteranomals (M weak) want more green in the mixture. Dichromats (i.e. protanopes (L blind) or deuteranopes (M blind)) accept all (red + green) mixtures as matching the

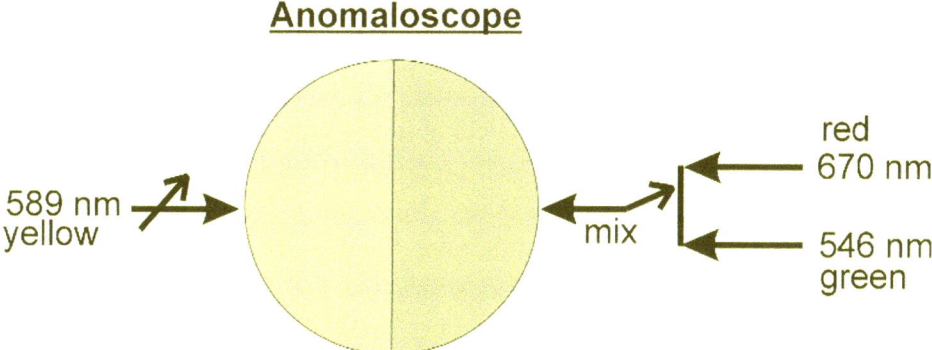

Anomaloscope

589 nm yellow → | ← **mix** ← **red 670 nm** / ← **546 nm green**

Match yellow (free choice intensity)
to mix-% red vs. green

Type colour vision	Anomaloscope-setting for match to yellow
normal trichromat	*limited variation %red matched to yellow*
anomalous trichromat	*mostly larger variation %red matched to yellow*
- deuteranomal	less %red than normal for match to yellow
- protanomal	more %red than normal for match to yellow
dichromat	*any %red matched to yellow*
- deuteranope	*c.* normal intensity yellow for match to high %red
- protanope	very low intensity yellow for match to high %red

Plate A.2 Diagnostic possibilities using an anomaloscope. A black and white version of this figure will appear in some formats. For the colour version, please refer to the plate section.

yellow half-circle, provided that the yellow intensity is properly adjusted; this is true from 100% red to 100% green. For the protanopes (L blind), red is then matched to a much darker yellow than is the case for deuteranopes (M blind).

An anomaloscope is needed for a clear distinction between dichromats and marked cases of anomalous trichromacy. Furthermore, anomaloscope measurements are very suitable for the quantification of the degree of disturbance in anomalous dichromacy (e.g. the range of acceptable (red + green) mixtures). In classical versions of the anomaloscope, such as the Nagel instrument, the yellow reference colour was obtained from a sodium lamp (monospectral light of 589 nm), the red from a lithium lamp (monospectral 670 nm), and green from a mercury lamp (monospectral 546 nm). Additive mixtures of the red and green lights were obtained using rotating prisms. In a classical Nagel anomaloscope, the scale for the mixing ratios (i.e. for the adjustment knob) is such that the pure green = 0 and pure red = 73. Thus, while dichromats accept all values from 0 to 73 as yellow (with adjusted intensity), normal trichromats typically require mixture readings only of between about 40–48. In a large dataset from London,[219] the ranges of yellow-matching (red + green) mixtures were 40–70 for protanomals (L weak), 0–45 for deuteranomals (M weak), and 0–65 for a particular group of 'extreme anomals'. Thus, for a few of the anomalous trichromats even the mixtures of normal trichromats were acceptable, but within a much wider range of acceptable variation. For another minority among the anomalous trichromats, the range of acceptable (red + green) mixtures was as narrow as those for normal trichromats but their mixtures had other, non-normal proportions. The width of acceptable variation in matching (red + green) mixtures varied enormously between different anomalous individuals, from about 5 or lower up to about 60 or more. For a range of acceptable (red + green) mixtures exceeding about 10, colour discrimination is usually deemed to be weaker than normal.

Modern anomaloscopes typically use various kinds of narrow optical filters rather than monochromatic light sources, and the measurement results are usually sent directly to a computer for further processing. Furthermore, besides the Rayleigh match used for investigating red-green blindness, there is also an analogous kind of match used in cases of disturbances in blue-green discrimination (*Moreland match*). Unfortunately, these highly useful instruments tend to be very expensive and they are therefore typically available only within laboratories and departments involved in research projects concerning normal or disturbed colour vision. As the classical anomaloscope requires monochromatic light sources, its functions cannot be reliably imitated using computers and their RGB monitors.

A.5 The neutral point of dichromats

As anomaloscopes are so expensive, it is interesting to consider whether some simpler procedure might be used for determining whether a red-green blind person is a dichromat or an anomalous trichromat. One theoretical possibility, that might present difficulties in practical use, would be to find out whether the participant has a truly 'neutral' point in a sunlight spectrum, i.e. a narrow region lacking chromatic colour. For red-green blind dichromats, this region would be located close to the spectral transition between blue and green (see 'NP' in Plate 5.2d, e).[220] Interestingly, the colour of this region is close to that for the most common variety of green traffic lights, which are therefore often looking grey/white for a dichromat.[221] In principle, a truly 'neutral' spectral point should only be present for dichromats and not for anomalous trichromats; however, in practice a clear distinction might be difficult to make because for many anomalous trichromats the colours are rather unsaturated in the blue-green region of the spectrum (Figure 5.4).

A.6 Genetic tests

In a not too distant future, it will probably become practically feasible to use molecular techniques of DNA analysis for determining whether somebody has an inherited kind of colour blindness and, if so, of which type and degree. Thanks to modern genetic techniques, it recently became possible to determine the kind of colour vision deviation for the long-dead eyes of Dalton: his genes demonstrated that his eyes were those of a deuteranope, lacking functional M cones.[222] A DNA analysis concerning colour blindness might be done with cell-material from any tissue in the body, e.g. from blood samples or from cells collected from the mouth with buccal swabs. In relation to colour vision, such techniques are already used in research laboratories but they are not yet sufficiently developed for routine applications on a massive scale.[223] Using DNA analysis, the present methods may identify 100% of the inherited deviations in L-cone visual pigments, but only about 65% of those for the M cones. When these techniques have been further developed and evaluated, a screening/initial diagnosis of inherited colour vision deviations might be more easily and rapidly done using DNA analysis than with classical physiological procedures (e.g. Ishihara + D15 + anomaloscope).

Appendix B   Specification and measurement of colours

More or less quantitative coloristic specifications are important in many branches of industry, controlling whether various fabricated items have the specified and/or required colour. Such determinations might concern, for instance, consumer articles like food, cosmetics, textiles, clothes, furniture, car seats, etc. There are at least three alternative procedures for the specification of colours: (1) direct comparisons to samples from various kinds of colour charts; (2) measurements using a (tri-stimulus) colorimeter; (3) measurements using a spectrophotometer.

B.1 Colour charts and systems

In the first chapter of this book, I dealt with the rather limited everyday vocabulary used for describing the confusing multitude of possible colours (Sections 1.1, 1.2; Plate 1.2). However, it is sometimes necessary to be able to use a more differentiated terminology for defining precisely which colours one sees or wants to see. As a basis for such definitions, the colours need somehow to be ordered into a system.

From ancient times, various philosophers and scientists have tried to analyse how all the many colours arise and how they are linked and interacting. In the course of history, hundreds of such surveys have been assembled; more than 160 schemes of this kind were summarized and reviewed in a recently published book.[224] Such lists and overviews have been made with different main purposes in mind:

1. *Analytical surveys*: colours ordered in relation to scientific findings concerning the physical/physiological basis for colour differences. This includes the systematic ordering of colours as seen in the solar spectrum (Plates 2.1b, 2.4), often arranged in circles with complementary colours opposite to each other (Plates 2.1c, 3.4b, 3.5b). Such analytical schemes form the basis for many of the more practically oriented surveys.
2. *Practical surveys*: lists and schemes providing easily used definitions for great numbers of different colours (Section B.1.1).
3. *Measurement schemes*: quantitative systems for the graphical analysis and summary of measured colour parameters (Section B.1.2).

These various strategies are not mutually exclusive, and several colour maps have been made with more than one of these purposes in mind. Besides systematic surveys of the kinds 1–3, there are also other kinds of summaries in which colours are ordered according to various aesthetic, emotional or other subject-ive/introspective aspects. Examples of such popular summaries are continually published in great numbers of books and articles, e.g. concerning clothes, gardens, home decoration, religion and spirituality, etc. Such summaries and surveys will not be further considered below (for further comments, see Section 1.5).

B.1.1 Colour systems for practical use: Munsell colour system

Colour systems within which all the usable and possible colours may be quickly and precisely defined are of great practical importance, e.g. for buyers and sellers of various kinds of paint, for industrial applications, etc. Furthermore, compre-hensive systems for colour identification and naming have long been important in various scientific contexts, e.g. for the description and classification of animals, plants and minerals. Such colour systems usually include a great number of representative colour samples which are ordered and encoded in ways that are intuitively easy to learn and apply. Ideally, the samples of such systems should be chosen such that the step-wise differences between neighbouring colour shades are perceived as being about equally large. There are very many such systems, typically constructed along the three major colour dimen-sions of hue, colourfulness (saturation) and brightness (cf. Plate 1.1). One of the first three-dimensional systems to be published was a beautiful *Farbenkugel* (1810),[225] designed by the German artist P. O. Runge (1777–1810). A well-known modern example is the Munsell system (Plate B.1),[226] here described as a general representative for such sample-oriented colour systems.

The American artist Albert H. Munsell (1858–1918) published an early version of his colour system in the book *A Color Notation* (1905). His system was the first one to use a consistent subdivision of the colour specifications into three independent dimensions corresponding to *hue, saturation* (also called *chroma*) and *brightness* (also called *value*) (cf. Plate 1.1)[227] There were several earlier three-dimensional systems, but none with such a strict distinction between, in particular, brightness and degree of saturation. The base of the Munsell system is a colour circle whose periphery is subdivided into five equal parts for the hues of red, yellow, green, blue and purple. Each one of these circular fifths is further subdivided into 20 equal fields, showing steps of gradual shifts between the local main hue and its neighbours, i.e. varying nuances of yellow-green, red-yellow, etc. Thus, the whole colour circle contains a total of 100 different

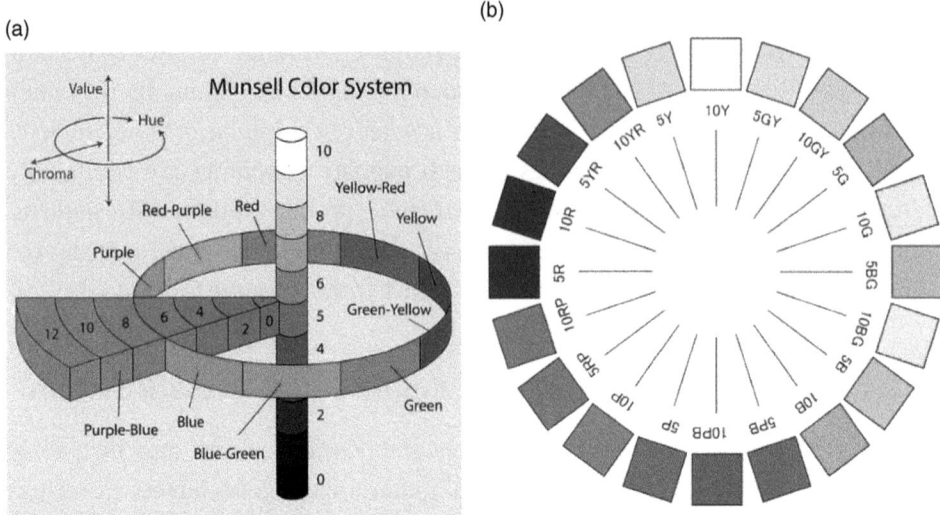

Plate B.1 The Munsell system for the practical organization of colours. The three dimensions of the system are called 'hue', 'chroma' (i.e. level of colourfulness or saturation) and 'value' (i.e. relative brightness or lightness). The hues are organized in a circle (a, b) with the most saturated colours along the rim and progressively less saturated colours toward the centre. At the centre, colours are achromatic (white/grey/black; a). Colours of different value (i.e. 'lightness') are represented in similar circles, vertically layered above each other (at centre: black at the bottom, white at the top). Source: (a) http://commons.wikimedia.org/wiki/File:Munsell-system.svg. Image by Jacob Rus. (b) http://commons.wikimedia.org/wiki/File:MunsellColorWheel.png. Image by PlusMinus. A black and white version of this figure will appear in some formats. For the colour version, please refer to the plate section.

saturated hues. Complementary colours are shown opposite each other, i.e. colours for which an appropriate additive mixture gives grey/white. Grey/white is also indeed shown at the centre of the circle (cf. Newton's circle, Plate 2.1c), and from the periphery toward the centre the colour of each particular hue becomes progressively less saturated while maintaining the same lightness. In the three-dimensional version of this system, colour circles of different brightness are placed on top of each other, thus creating a 'colour solid' analogous to Runge's *Farbenkugel*. In a Munsell colour solid, the achromatic centre is white at the top and black at the bottom with all shades of grey in between. In such a three-dimensional system, each possible colour may be precisely encoded using only three coordinates, and the position of each colour is relatively easily retrieved once the principles behind the system architecture are understood. In practice, the Munsell system is used together with colour samples from various parts of the colour solid. Such samples are, for instance, illustrated in the *Munsell Book of Color*; the most comprehensive edition of this publication includes examples of more than 1600 colours.[228]

Other modern and internationally used systems with a similar range of practical applications include the Swedish Natural Color System (NCS),[229] the German DIN system (Deutsches Institut für Normung) and the two American systems OSA UCS (Optical Society of America's Uniform Color Scales), and Pantone (Pantone Matching System, PMS).[230]

B.1.2 Measurement schemes: CIE chromaticity charts

In colour science, a major problem concerns the prediction of perceived colour when knowing only the physical properties of the light entering the eye, i.e. its intensity at different wavelengths. For a small perceptual field, each intensity/ wavelength pattern produces a predictable colour perception, but the same perceived colour may be evoked by many different wavelength combinations (*metameric* colours, cf. Plate 2.11). This is due to the fact that we do not directly perceive the wavelength pattern of light, but our colour perception is dependent on the relative degrees of activation of three different sensors, each one having its own characteristic distribution of wavelength sensitivity (cf. Plate 4.3a). The straightforward relation between perceived colour and single wavelengths along a solar spectrum does not inform us as to the effects of various *wavelength mixtures*. For such purposes, *chromaticity diagrams* are highly useful. These diagrams are most directly applicable for direct additive mixtures of light. In everyday situations concerning reflected light, the source of illumination and the reflectance of the target material have to be considered as well.

The chromaticity diagrams of the International Commission on Illumination (Commission Internationale de l'Eclairage, CIE) were briefly introduced in Chapter 2 (Section 2.5; Plate 2.6). The CIE chromaticity charts are based on empirical psychophysical measurements in which matches were investigated between a target colour and an additive mixture of three primary colours: red, green and blue. The participant would see the target colour and the tri-stimulus mixture side by side (cf. similar method in an anomaloscope, Plate A.2), within a small isolated portion of his/her central field of vision (originally 2°). The effects of the tri-stimulus mixture of primary lights were expressed using three fictive light sensors (Plate B.2a), each one with a different sensitivity for different wavelengths (for methods of calculation, see below, Section B.1.2.2). The chart resulting from such measurements and calculations (Plate B.2b, c) combines two

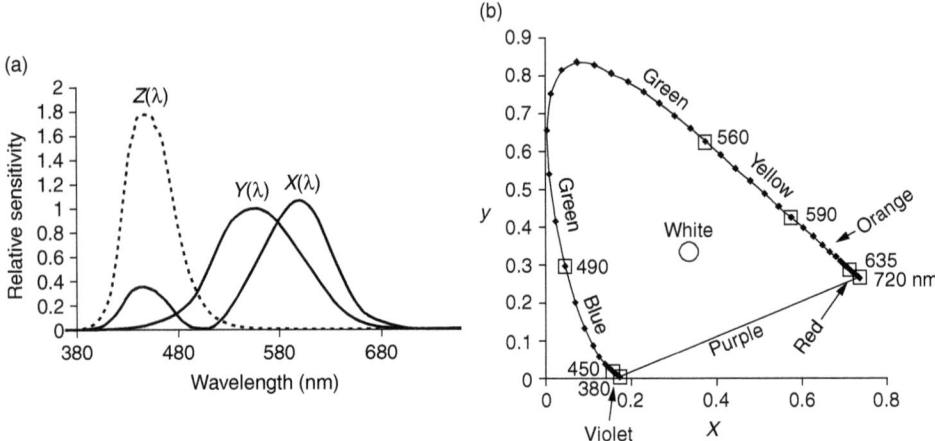

Plate B.2 (a) Sensitivities to light of different wavelengths for the three imaginary sensors (cf. human cones) of the 1931 CIE 'standard observer', used for calculating the CIE chromaticity curve. **(b)** Outer rim of the chromaticity curve, representing the maximally saturated colours. Values calculated from the responses of the imaginary 'standard observer' to monochromatic light of different wavelengths (see Section B.1.2.2 for calculations). In this diagram, approximate wavelength-regions (nm) for different colours are indicated: 380–450 violet, 450–490 blue, 490–560, green, 560–590 yellow, 590–635 orange, >635 red.
Using data from http://www.cvrl.org: CIE 1931 2-deg, XYZ CMFs.

(c)

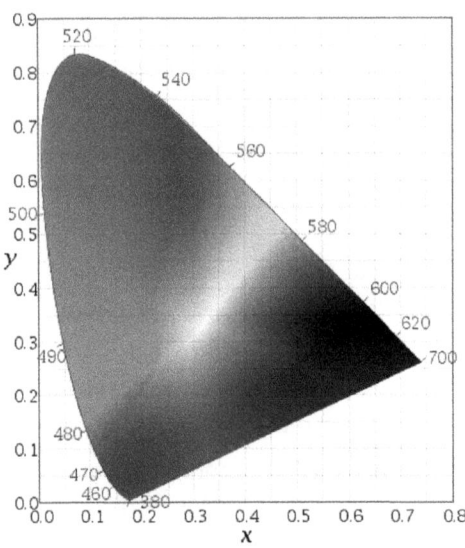

(c) Approximate sites of different colours within the CIE chart, including the less saturated colours lying inside the rim of the chromaticity curve.
Source: http://en.wikipedia.org/wiki/File:CIExy1931_fixed.svg.
Image by Sakurambo.
A black and white version of this figure will appear in some formats. For the colour version, please refer to the plate section.

different sets of data: (1) a coordinate system for the relative tri-stimulus values (xy values shown directly, z values implied, i.e. equal to $(1 - x - y)$), and (ii) indications of the colours actually matched and perceived for each one of these activation patterns (i.e. shown directly as colouring or as text labels). The horseshoe-shaped curve in the CIE chart shows lines connecting the xy-para-meters for monospectral light stimuli, covering the whole visible spectral range of light, from violet at 380 nm via blue, green, yellow and orange to red at 700 nm or more. The 380 nm (violet) and 700 nm (red) corners are connected with a straight line, representing light stimuli evoking purple hues. Thus, the horseshoe-shaped CIE curve shows the xy-activation parameters that will elicit maximally saturated colours. Inside the curve are all the less saturated colours, and at the centre of the inside space, at $x = y = 0.333$, are the achromatic colours (i.e. black, grey, white). In many of its general properties, the CIE chart resembles the colour circle of Newton (Plate 2.1c). A single CIE chart, like that of Plate B.2c, represents activation patterns that will produce colours with the same brightness but with different *chromaticity*, i.e. colours differing in hue and saturation. In the CIE system, the remaining colour parameter, the brightness or luminosity (cf. Plate 1.1; here labelled parameter Y), is represented by a vertical axis of variation. Thus, chromaticity charts of different brightness are thought of as being stacked on top of each other with dark ones at the bottom (achromatic centre black) and progressively brighter ones further up (achromatic centre white at the top; cf. three-dimensional shape for the Munsell system, Plate B.1a).

   The CIE chromaticity chart is constructed such that it might easily be used for predicting the results of any additive colour mixture. For instance, for a mixture of two colours, all the resulting colours are lying on the straight line drawn between the activation patterns of the two starting colours. If this straight line passes through the achromatic centre of the chart (Plate B.2b, 'white'), this means that the two starting colours are complementary, i.e. capable of being mixed such as to give grey/white.

B.1.2.1 Nomenclature for specifying colours in xyz CIE chromaticity charts

In the CIE chromaticity diagram, the hue of an unsaturated colour is calculated by drawing a straight line from the central white point, passing through the x/y site of the target colour, and continuing until the line cuts through the saturated rim of the CIE curve (Figure B.1). The hue of the unsaturated target colour, i.e. its *dominant wavelength*, corresponds to the spectral colour at the intersection between the straight line and the rim of the CIE curve. Thus, in Figure B.1 the colour a has a yellow hue corresponding to the colour of

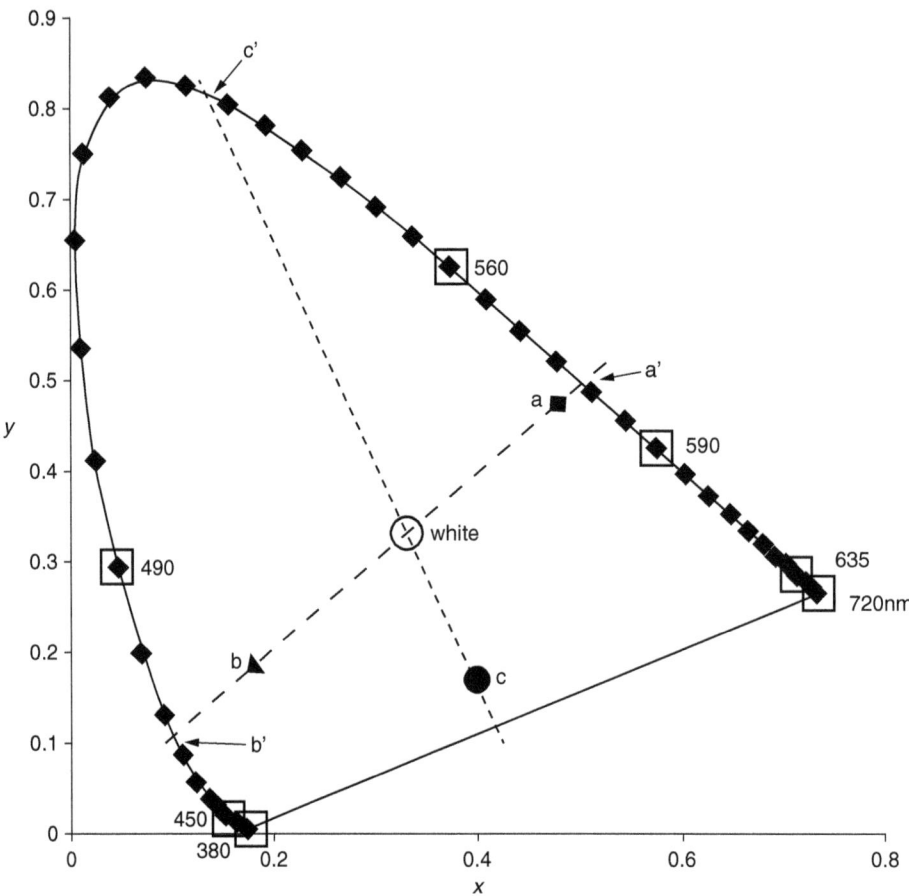

Figure B.1 CIE-1931 chromaticity chart (cf. Plate B.2 b), showing examples of how this system may be used for determining the dominant hue and wavelength of a colour. The two complementary target colours a and b correspond to the yellow and the blue of the Swedish flag. In each case, a straight evaluation line has been drawn from the central white point, through the respective colour point (a or b) and further out until the periphery of the chromaticity curve is transected, i.e. the curved line representing the sequence of spectral colours and their wavelengths. The point of transection between the straight evaluation line and the spectral curve defines the *dominant wavelength* for the target colour. In the example this transection lies at a' ~579 nm for target colour a (yellow) and at b' ~477 nm for target colour b (blue). For purples, e.g. target colour c, the straight evaluation-line is drawn through the white point and then further on until the spectral curve is transected at a complementary colour of the target. For target colour c, this *complementary dominant wavelength* corresponds to a green light of about 528 nm.

monospectral light of 579 nm (see a' and thin interrupted line). In the case where the straight line would intersect the CIE curve in its purple extra-spectral section, the hue of the target colour is instead, somewhat confusingly, defined by its complementary value. Thus, for the purple target colour c in Figure B.1, the *complementary dominant wavelength* is about 528 nm, which corresponds to a greenish hue (see c' and interrupted line).

The various versions of the CIE chromaticity system are very useful and much used for the unambiguous definition of colours. For instance, in the CIE 1931 version, the colours of the Swedish flag are defined as corresponding to $x = 0.472$; $y = 0.465$; $Y = 64.4$ for the yellow cross and $x = 0.189$; $y = 0.192$; $Y = 8.3$ for the blue background.[231] Y here signals the relative lightness (reflectance) of the respective colours; thus, the numbers indicate that the flag's yellow is much lighter than the blue. In Figure B.1, the x/y positions for these flag colours are labelled a and b for the yellow and the blue.

B.1.2.2 Calculations using the 1931 xyz CIE chromaticity chart

The coordinates of a colour in the CIE chromaticity chart may be calculated from measurements of the intensity of the stimulating light at different wavelengths, i.e. from spectrographic data (e.g. Plates 2.12c, 2.13b, c). For the 1931 CIE chart, these calculations are done using the three fictive colour sensors of the so-called 'standard observer'. These sensors are called $X(\lambda)$, $Y(\lambda)$ and $Z(\lambda)$, and their respective relative sensitivities for different wavelengths of light are shown in Plate B.2a. With these sensitivity profiles, the degree of light-induced activation may be calculated for each sensor type, using spectrographic intensity/wavelength measurements. For the graphic display of an *xyz* chart, the respective absolute activation levels (X, Y and Z) are transformed into relative plot-values (x, y and z) using the following simple equations:

$$x = X/(X + Y + Z) \tag{B.1}$$

$$y = Y/(X + Y + Z) \tag{B.2}$$

$$z = 1 - (x + y) \tag{B.3}$$

The Y parameter has a double function in the xyz CIE chart: besides being used for the normalized *xyz* coordinate calculations, its value also provides a measure for colour brightness. In the 1931 *xyz* CIE chart, only the x and the y values are plotted directly; the z values may be deduced from the x and y values using Equation (B.3). Considering the sensitivity profiles of the fictive colour sensors, longwave red colours would be expected to have a high x-value, middle-wave green colours would have a high y-value, and shortwave blue/violet colours would have high z values, i.e. low values for both x and y. In accordance with these principles of the CIE chart construction, the perceived colours are indeed green at the top (high y), red at lower right (high x) and blue/violet at the lower-left corner (low x and y).

The three sensor functions of Plate B.2a and the resulting chromaticity diagram (e.g. Plate B.2b) represent the final mathematically adapted 1931 version of the CIE model. In this CIE model, the wavelength sensitivity of the $Y(\lambda)$ and $Z(\lambda)$ sensors resemble those of the M and S cones, while the wavelength sensitivity of the $X(\lambda)$-sensors has another distribution and shape (bimodal versus unimodal) than that usually shown for the retinal L cones (cf. Plates B.2a and 4.3a).[232] The lack of a direct and complete correspondence between the sensors of the psycho-physical CIE 1931-model (Plate B.2a) and the biological sensors of the retina (Plate 4.3) is partly for technical reasons.[233] Despite such apparent differences between model and biology, the 1931 version of the CIE chromaticity diagram is still very useful and widely used.

B.1.2.3 Different versions of CIE chromaticity charts: other colour spaces

For the 1931 version of the CIE *xyz* diagram (*CIE 1931 Color Space*),[234] the psychophysical measurements were done for a narrow central portion of the visual field, covering only 2° (retinal fovea; cf. Figure 3.2). There are several other CIE model versions, including an *xyz* model from 1964 which is valid for a larger visual field of 10°.

One of the disadvantages of the classical 1931 *xyz* model is that, within the chromaticity diagram, the physical distances between the colours do not give a linear quantitative representation of their perceptual differences. Thus, for instance, a disproportionately large upper section of the 1931 *xyz* diagram is occupied by a range of rather similar greenish-yellowish colours (Plate B.2c). These aspects of the model have been improved in another much used three-dimensional version called the 'Lab color space', in which the classical *xyz* coordinates have been transformed using appropriate formulas.[235] Originally, this model version was called $L^*a^*b^*$ according to its three coordinate axes, in which L^* corresponds to lightness (0 for black, 100 for white), and the a^* and b^* values concern positions along the respective lines between two pairs of opponent colours: a^* from negative values for green toward positive values for purple (magenta), b^* from negative blue values to positive yellow values.

Using mathematical formulas, values of the CIE 1931 colour space may also be converted to those for various non-CIE color spaces, and vice versa. One of several alternatives is the *cone colour space* (or *LMS colour space*), based on the sensitivity characteristics of the human L, M and S cones.[236] This colour space is, for instance, often used when estimating the appearance of a sample under different illuminants.

B.2 Measurements using a colorimeter

Colorimeters quantify colours by measuring the composition of light reflected from an object when illuminated by white light (e.g. sunlight). The classical kind of colorimeter, also called a *tri-stimulus colorimeter*, is an instrument containing three sensors for light intensity, each one provided with its own optical filter. The filters limit the wavelength sensitivities of the three sensors such as to make them correspond to the three sensors of the CIE 'standard observer' (Plate B.2a). Due to this construction, the measurements of the colorimeter are very easily converted to colour coordinates of the type used in the CIE chromaticity diagrams (Section B.1.2).

B.3 Measurements using a spectrophotometer

Spectrophotometers may be used for quantifying colours as well as for various other tasks. Such an instrument contains tens of light intensity sensors, each one provided with an optical filter of narrow bandwidth, allowing the passage of only a limited range of wavelengths. For instance, for the measurements of Plates 2.12c, 2.13b, c and Figure 5.5, a spectrophotometer was used with 36 different sensors, measuring the light intensity at points 10 nm apart for light of 380–730 nm. Also from such measurements, the CIE chromaticity coordinates might be calculated and, hence, the colour defined (Section B.1.2).

As compared to the tri-stimulus colorimeter, the spectrophotometer provides results that are more detailed and of more general usefulness. For instance, using a spectrophotometer, differences might be analysed between metameric colours, i.e. colours that differ in wavelength composition but look the same for human eyes and tri-stimulus colorimeters (Plate 2.11).

Besides their use in colour analysis, spectrophotometers are very important and commonly used for various kinds of chemical analysis. For instance, the content of a chemical compound in a liquid solution might, for many substances, be analysed by measuring how the different wavelengths are absorbed as light is passing through the solution. Similarly, astronomers use spectrophotometers for determining the chemical composition of the atmosphere surrounding distant planets and stars.

Appendix C   Light and lighting

C.1 Wavelength and oscillation frequency of light

There is an inverse proportional relation between the wavelength (λ) and frequency (f) of light according to the formula: $f = c/\lambda$, in which c is the speed of light in a vacuum (about 300 000 km/s). For visible light, wavelength is usually measured in nanometres (nm) and the frequency in terahertz (THz). There are 10^9 nm per metre (i.e. 1000 million) and a frequency of 1 THz means that there are 10^{12} oscillations per second (i.e. a million million). The oscillation frequency of light in terahertz is obtained by dividing 300 000 by the wavelength in nanometres. Visible light has wavelengths of about 380–730 nm, which thus corresponds to oscillation frequencies of (300 000/380) to (300 000/730) = 789 to 411 THz (cf. Plate 2.4).

C.2 Light refraction

The famous speed of light of about 300 000 km/s is only valid in a vacuum, and it is then alike for all wavelengths. However, when passing through various materials the speed of light goes down, and this slowing is more marked for shortwave than for more longwave light. This slowing-down causes a light beam to change its direction (to refract) when passing an oblique borderline between two different translucent materials, e.g. at the border between air and glass (or air and water). The refraction is greater the shorter the wavelength. Hence, as was investigated by Newton in his famous experiment, light may be separated into differently coloured components by letting it pass through a glass prism (Plate 2.1b). The degree of slowing-down and the resulting refraction is different for different materials, and a measure of this property is given by the index of refraction, which is equal to the ratio between the speed of light in a vacuum and the speed in the target material. For instance, the index of refraction is 1.0003 for air, 1.33 for water, between about 1.47 and 2.04 for different kinds of glass, and unusually high for a glittering material like diamond (2.417).

C.3 Quality of lighting

Practically everything we see in the outside world depends on light reflected from various objects. For colour vision, both the quality of the light (its wavelength composition) and its intensity are of great importance. In very weak light we see no colours because we are then using only our rods. In stronger light, the wavelength composition of the illumination may have great effects on the apparent colour of a given object (cf. Section C.4). For instance, with a strong monochromatic illumination, like the yellow light from sodium lamps, no colour distinctions would be possible because truly monochromatic light has only one wavelength and one colour.

In the spectrum of midday sunlight, the intensities are rather similar for the different visible wavelengths, with a peak at around 500 nm (Figure 2.2).

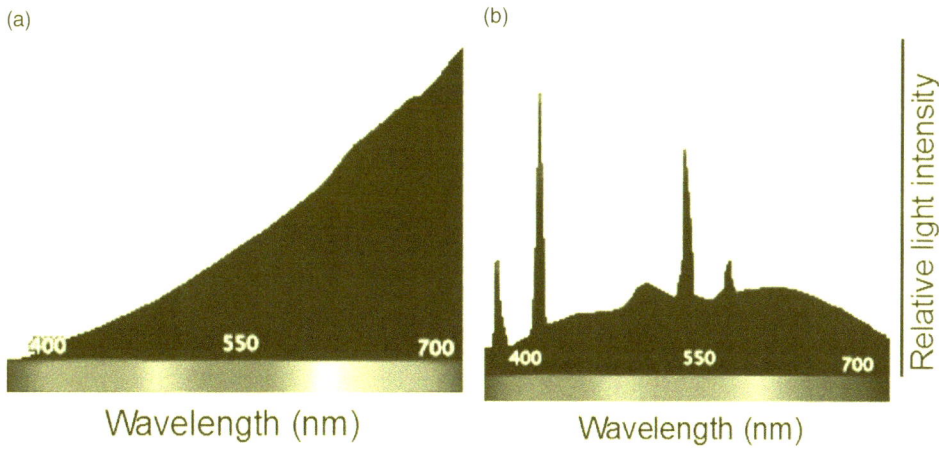

Plate C.1 Examples demonstrating the differences in wavelength composition between the light produced by incandescent light bulbs (a) and by fluorescent lamps (b). See section C.3 for comments.
Source: http://en.wikipedia.org/wiki/File:Spectral_Power_Distributions.png.
Image by Dkroll2.
A black and white version of this figure will appear in some formats. For the colour version, please refer to the plate section.

This peak corresponds to the radiation obtained from a 'black body' at a temperature of 5777K (Kelvin degrees = Celsius + 273.15). At higher temperatures, this thermal radiation becomes more energetic and dominated by shorter wavelengths, and for lower temperatures the opposite change occurs. In old-fashioned incandescent light bulbs, the light is produced by thermal radiation at much lower temperatures than that of the sun; hence, the illumination produced by such light bulbs is dominated by longer wavelengths than those of sunlight (cf. Plate C.1a). It would be technically difficult and cost a lot of energy to have light bulbs glowing at sun-like temperatures; therefore, other techniques have been employed for the artificial production of a more sun-like and colour-friendly illumination. This has mainly been done using various kinds of fluorescence (Plate C.1b) or electroluminescence (e.g. LEDs). By such means, illuminating light with a similar intensity across the various wavelengths may be produced, which makes reflected colours look rather similar to those seen in daylight. This lighting has the added advantage of being less energy demanding than the light from incandescent light bulbs. However, inside the lamp, the fluorescence is typically activated using UV light, which is emitted by electrical activation of mercury gas. This activating 'mercury radiation' contains narrow peaks of intensity at certain wavelengths (Plate 2.10), and these peaks will also, of course, appear in the total light emission of the fluorescent lamp (Plate C.1b). Such narrow peaks of increased light intensity might cause some technical problems in light measurements.

C.4 Lighting and metameric colours

For all types of colour reproduction, the combination of wavelengths used for each colour is likely to be quite different in the reproduction as compared to in the original. The main aim of a colour reproduction is not to imitate the distribution of wavelengths found in the original, but to produce a copy capable of causing the same human colour perception. Colours looking precisely the same might be evoked by very many different wavelength combinations; this phenomenon is called *metamerism* (Plate 2.11). For reflected colours, the metamerism is highly dependent on the composition of the illuminating light. For instance, two objects of seemingly the same colour in daylight or in fluorescent lamp light might look quite different from each other in incandescent lamp light (cf. wavelength distributions in Plate C.1a versus b).

The risk that disturbing colour changes might occur with a change of illumination is particularly great if an achromatic target colour (e.g. grey) is being produced by the mixture of a limited number of complementary chromatic colours (e.g. yellow + blue). If one of the complementary mixture colours is not well represented in the illuminating light (e.g. rather little blue in incandescent bulb light), then the grey object will stop looking grey.

C.5 Total intensity of lighting

When quantifying the total intensity of illumination, the measurements might concern the intensity at the light source, the intensity across the surface that becomes illuminated, and the intensity of light arriving at the eye. In the SI-system, the intensity of the light source is measured in candelas (cd), which is the Latin word for a candle. Accordingly, the intensity of 1 cd does indeed correspond to the approximate intensity of light emitted by a common candle. From such a light source of 1 cd, a light flow of 1 lumen (lm) per steradian (sr) is emitted in all directions. The intensity of *illumination* of a surface area is measured in units of lux (lx), and 1 lx = 1 lm/m². For the reflected light reaching the eye, the *luminance* (L) is measured in units of cd/m², sometimes referred to as 'nits'. The luminance may be calculated from the *illumination* (E), provided that the illuminated surface reflects the light in all directions (i.e. has no mirror properties). The formula used for such calculations is:

$$L = \beta(E/\pi) \tag{C.1}$$

with β equal to the reflectance of the illuminated surface (e.g. about 0.8 for white paper).[237] Thus, white paper illuminated by 1000 lx gives a luminance of 255 cd/m².

In order to give the reader a more concrete impression of what these complex units mean, Table C.1 shows the intensity of illumination (lx) under different circumstances, and the intensity of light flow (lm) from different light bulbs.[238] Cones remain functional down to a luminance of about

Table C.1 Measurements of illumination under different conditions

Outside illumination	
0.000 01 lx	Threshold for scotopic vision (i.e. vision using rods)
0.000 1 lx	Cloudy night without moon
0.001 lx	Clear night, stars but no moon
0.01 lx	Quarter moon
0.27 lx	Clear night with full moon
1 lx	Candle at 1 m distance
10 lx	Dusk (~maximum for useful rods)
100 lx	Very dark and cloudy day
400 lx	Sunrise or sunset, no clouds
1000 lx	Cloudy day
10 000–25 000 lx	Full daylight, indirect sunlight
32 000–100 000 lx	Full daylight, direct sunlight
Inside illumination, typical values	
50 lx	Living room
320–500 lx	Office
Light flow from incandescent bulbs	
25 W	200 lm
60 W	850 lm
75 W	1200 lm
100 W	1700 lm

0.001 cd/m^2, provided that the pupil has a diameter of at least 7 mm. Rods begin to become over-saturated (i.e. less sensitive for light intensity differences) at a luminance exceeding 3 cd/m^2, and they are insensitive to light differences at values exceeding about 10–20 cd/m^2; the more narrow the pupil, the higher is the maximal luminance intensity with a maintained rod function.[239] Hence, a simultaneous use of cones and rods (*mesopic vision*) is possible for luminance values between about 0.001 and 3–10 cd/m^2, i.e. for illumination values of about 0.004 to 12–40 lx (weakly moonlit night to dusk). The table shows that the illumination in full daylight may be up to 370 000 times stronger than the illumination at full moon; in both situations we can move around using our eyes, i.e. our eyes are marvellously capable of doing their job under extremely different light conditions.

Appendix D   Digital cameras

Our eyes are often said to be 'camera-like' (or, rather, our cameras are 'eye-like'): using a lens, a picture of the surrounding world is projected onto the reverse side of the eye/camera, where it is somehow recorded and retained for later use. In our eyes, the recording and further processing happens using receptors and nerve cells. In analogue cameras, the recording is done using a light-sensitive film. In digital cameras, the film is replaced by a light-sensitive electronic sensor and the further processing is done using computer techniques. There are many interesting similarities and differences between our visual system (eyes and brain) and the functional properties of modern digital cameras. In this Appendix, I will concentrate on colour-associated aspects.

D.1 Sensor and colour analysis

The light sensor of a digital camera contains a huge number of light-sensitive picture elements (*pixels*), often about 10 million or more. There are three types of elements, each one reacting to a different range of wavelengths: R pixels for red, G pixels for green, B pixels for blue/violet; this wavelength selectivity is typically produced by placing a different optical filter on top of each kind of sensor element. The signals from the R, G and B pixels are analysed using an RGB model (cf. Plate 2.6). With methods similar to those of the CIE system, the RGB-signals may be used to calculate the colour to be produced in different portions of the picture.[240] The RGB-models are directly derived from studies of our visual system, and these digital camera techniques may indeed seem to be very eye-like. However, there are also several evident differences between eyes and the RGB-organization of digital cameras.

Camera sensors often have their RGB-pixels arranged according to a so-called *Bayer pattern* (Plate D.1) with 25% R, 50% G and 25% B pixels; the name of this pattern comes from Bryce Bayer, who worked at Kodak in the 1970s. This pixel pattern gives the camera sensor a maximal light sensitivity within the green region of the spectrum (cf. eye sensitivity 'In daylight', Figure 4.4). However, in the retina the proportions of the various light-sensitive elements are quite different from those of the Bayer pattern (Section 4.3.1); for instance, the eye has only about 5% S cones (cf. 25% B pixels). Furthermore, the spatial distribution of the retinal cones is very heterogeneous and complex (Section 4.3.2).

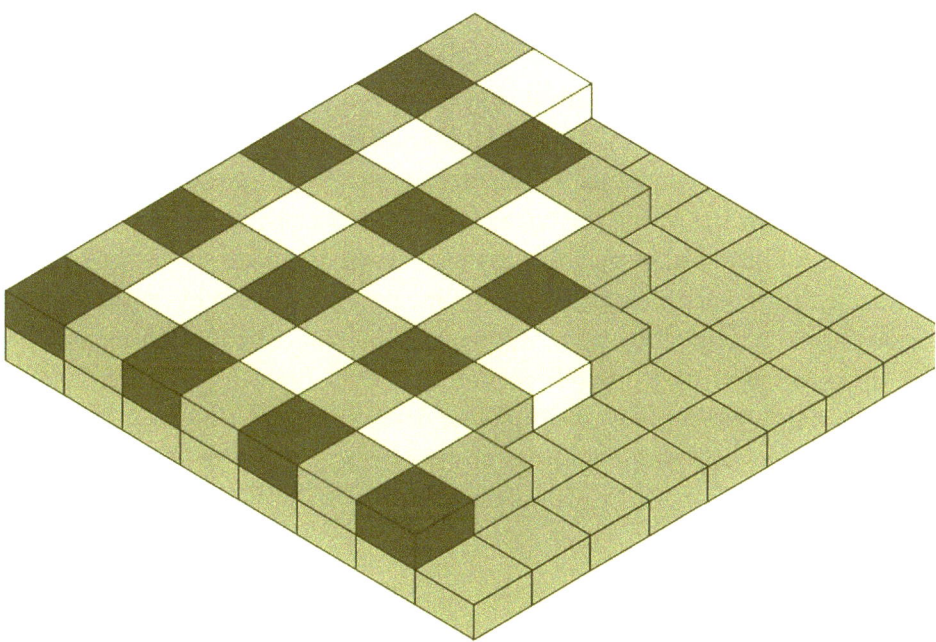

Plate D.1 Part of the light-sensor of a digital camera (schematic): the commonly used Bayer pattern for the distribution of pixels with different colour (i.e. wavelength) sensitivities, i.e. the R, G and B pixels of the RGB model.

In the camera sensor, all types of pixels are used together to calculate lightness contrast; in the eye this is done using only the L and M cones. In both the eye and the camera sensor, the colour of each picture detail must be calculated using the reactions of several adjoining pixels with different wavelength sensitivities. In the most central portions of the retina, the distance involved might be rather large because of the central lack of S cones (Section 4.3.2). The eye/brain can see more highly saturated colours than those represented in RGB-models for common types of camera sensor (Plate 2.6). Lenses have inherent problems with chromatic aberration, i.e. details become blurred due to differences in refraction for light of different wavelengths. In the eye, such problems are diminished by using only part of the spectrum for the analysis of lightness contrast; only the longwave L and M cones are used for such purposes, not the S cones (Section 4.3.2). In cameras, sophisticated lens constructions and correcting software are used for reducing the effects of chromatic aberration.

For details shown in lightness contrast (i.e. in black and white), the maximal resolution of the camera's light sensor depends on pixel densities as well as on signal processing. Small compact cameras may have a pixel density that is even higher than corresponding figures for the eye. In the retina, the central parts of the fovea have about 15 million cones/cm^2, while a small compact camera with a total of 14 megapixels (MP) might have a density up to at least 50 million pixels/ cm^2. However, these numbers are smaller for high-quality cameras; in full-format mirror-reflex cameras pixel densities are also smaller than those for the central fovea (e.g. about 1.4–4.2 million pixels/cm^2 for cameras of 12–36 MP).

In the retina, the functional sensor-density decreases very rapidly with distance outside the fovea, and the same is true for detail resolution. This peripheral decline is mainly due to an increasing tendency for many cones or rods to be connected to each neurone, thus enhancing light sensitivity at the cost of resolution. With regard to colour, the eye has different powers of discrimination for different retinal regions (cf. Section 3.3.3). In these respects the camera sensor is much simpler: the conditions for detail resolution and colour discrimination are the same within all parts of the camera's 'visual field'.

D.2 Light sensitivity

In our eyes, the sensitivity of the receptor cells automatically becomes higher in darkness, mainly due to an increased concentration of light-sensitive visual pigment (*dark adaptation*, Section 4.5.2), i.e. the receptor mechanism itself becomes more sensitive. In addition, we have two types of sensor in the retina (somewhat analogous to using an old-fashioned camera with two kinds of film): cones for daytime colour vision, rods for nighttime vision in black and white. Due to the wide convergence of many rods onto each single neurone, the high sensitivity of nighttime vision is associated with a relatively coarse-grain perceived picture.

Also in some modern cameras, the light sensitivity may become somewhat increased by using more sensor elements per functional pixel. However, this is not typical, and from several points of view the 'dark adaptation' of the camera differs from that of the eyes. First, the camera normally has only one general type of sensor elements, with fixed light sensitivity and colour properties. Thus, a camera does not become colour blind or unsharp in darkness. If shot with an adequate exposure time, a nighttime photo does not resemble the nighttime image perceived by a human eye: the photo becomes far too detailed and shows far too many colours. Second, the dark adaptation of the camera is not based on a change in light sensitivity, but rather depends on a change in the amplification of signals from the light-sensitive elements; this is what happens with a change in the camera's ISO settings. The increased amplification will then also be applied to the general background noise of the picture, and this noise will be relatively greater the weaker the signal (i.e. lower signal/noise ratio). Therefore, the maximum useful light sensitivity of a digital camera depends on its noise levels at high

degrees of amplification (at high ISO values). For the same amount of amplification, noise levels are generally larger for high than for lower pixel densities.

D.3 Colour constancy and white balance

In Section 3.1.3, I described how the brain is capable of interpreting a given
combination of wavelengths quite differently depending on the general illumination: white paper looks white in strong outside daylight, in the reddish light at
sunset and with the yellowish light of old-fashioned light bulbs. This phenomenon
is called colour constancy and the associated change in colour processing occurs
very rapidly, much faster than the dark or light adaptation. Colour constancy
adaptations mainly reflect shifts in ongoing processing programs of the brain and
they are not primarily dependent on alterations in retinal receptor mechanisms.

   If you take a photograph at night in a room illuminated by incandescent light
bulbs and the camera operates with settings (or film) meant for daylight, then
the picture will look unnatural and far too yellow when regarded in daylight. In
the period of analogue photography, one had to use different films for incandescent illumination and for daylight. Alternatively, when using daylight films,
pictures taken indoors at night might be illuminated by flash, or one might
compensate for the lack of shortwave indoor light by using a blue filter in front
of the camera (i.e. blocking much of the yellow light). Digital cameras are
cleverer and, for each kind of illumination, they try to adjust their colour
processing such that white continues to look white. This is called '*white balance*'
(*WB*, Section 3.1.3), and the process is typically fully automatic: the camera
measures the general distribution of incoming wavelengths and tries to calculate
how to adjust the effects of colour mixtures such that white continues to look
white. Generally this works OK, particularly for light sources in which the
intensity changes only gradually across the various wavelengths (e.g. heat
radiation from sun or incandescent light bulbs; Plate C.1a). However, cameras
may have difficulties with their white balance when confronted with strange
and unexpected illumination. For most kinds of digital cameras, it is then
possible for the photographer to calibrate the white balance, e.g. by using a
piece of grey paper that is illuminated by the strange light, and 'telling' the
camera to keep displaying this paper as grey ('preset' white balance). Cameras
might, for instance, apparently have difficulties with automatic white balance if
the illuminating light has one or several sharp peaks of intensity at certain
wavelengths (e.g. fluorescent light, cf. Plate C.1b).

Appendix E Technical terms

achromat – person who is unable to see chromatic colour; might be inherited (rod achromat) or due to brain damage; state: achromatopsia

achromatic colour – colour lacking hue; i.e. white, grey or black

anomalous trichromat – person with three types of cones, one of which has deviant properties; needs three primary colours for admixture of all hues, although typically in non-normal proportions

blind spot – blind area enclosed within portions of the retina and visual field, exit site of the optic nerve (*optic disc*), hence lacking visual receptor cells

brightness – apparent intensity/amount of light in a visual percept (Plate 1.1)

chromatic colour – colours with a hue, i.e. all colours except white/grey/black; cf. *achromatic colour*

chromaticity – hue and level of saturation

colourfulness – relative degree of saturation or 'purity' of a chromatic colour (Plate 1.1)

complementary colours – pair of hues for which an additive mixture may become white; each spectral and extra-spectral hue has its own complementary colour

cones – class of receptor cells of the retina, used for *photopic vision* in normal daylight; three types with different wavelength sensitivity (L long, M middle, S short wavelengths); essential for seeing colours and fine detail

dark adaptation – adjustment of the sensitivity of visual receptor cells and the processing in the visual neuronal system to a decreased light intensity (e.g. from day to night)

deuteranomal – person with an inherited disturbance of red-green colour vision in which the M cones have deviant properties, state: deuteranomaly

deuteranope – person with an inherited disturbance of red-green colour vision in which the M cones are non-functional; state: deuteranopia

dichromat – person with two types of functional cones, needs only two primary colours for admixture of all hues; state: dichromacy

dominant wavelength – wavelength of spectral colour that represents the hue for an unsaturated colour; obtained using CIE charts (Figure B.1)

elementary colours – six colours perceived as 'unique' according to Hering's opponent colour system; four are chromatic (blue, yellow, green, red) and two achromatic (white, black)

extra-spectral colours – purple hues, do not occur in the spectrum because they cannot be produced using light of a single wavelength; made with e.g. blue + red

fovea – small retinal region for which acuity is maximal, lies at the centre of the normal gaze direction; corresponds to a small dimple in the retina

ganglion cells – retinal neurones which send axons to the brain via the optic nerve; have only indirect contacts with the receptor cells, via other nerve cells

green-blind/green-weak – see deuteranope/deuteranomal

hue – the colour property indicated by names like red, yellow, green, blue

L blind/L weak – see *protanope/protanomal*

L cones – see *cones*

LGN = *lateral geniculate nucleus*, a collection of nerve cells in the diencephalon which are the main recipients of signals (i.e. synaptic contacts) from axons of retinal ganglion cells. The LGN cells send their axons and signals to the primary visual cortex (= region V1 = area 17).

light – visible electromagnetic radiation, i.e. radiation at wavelengths that have an effect on the activity of visual receptor cells in the eye. These wavelengths cover a range of about 380–730 nm, corresponding to oscillation frequencies of about 789–411 THz.

light adaptation – adjustment of the sensitivity of visual receptor cells and the processing in the neuronal visual system to an increased light intensity (e.g. from night to day)

luminance – the luminous intensity of a surface in a given direction (Section C.5)

luminosity – lightness, or relative brightness, of something

M blind/M weak – see *deuteranope/deuteranomal*

M cones – see *cones*

macula lutea – see *yellow spot*

mesopic vision – vision at light intensities that allow a simultaneous use of rods and cones (e.g. with moonlight or candlelight)

metameric colours – colours that differ in wavelength composition but may look the same under suitable types of illumination

μm = micrometre = 10^{-6} m = 1 micron = 1/1000 mm

monochromat – individual with only one type of cone in the retina, cannot distinguish different hues (e.g. for humans: rare cases with only S cones)

monochromatic colour – hue produced by light of only one (or a very limited range of) wavelength(s); see *spectrum*

nm = nanometre = 10^{-9} m = 1/1000 μm

normal trichromat – individual with normal human colour vision, has three kinds of retinal cones, three primary colours needed for admixture of all spectral hues

opponent colours – complementary pair of *elementary colours*

opsin – protein component of visual pigments, determines the wavelength sensitivity of the pigment, different types for different categories of visual receptor cell, works together with *retinal*

optic disc – exit site of the *optic nerve* at the back of the eye, lacks retinal tissue and gives rise to the *blind spot*

optic nerve – bundle of nerve fibres connecting the eye to the brain; c. 1.2 million fibres per human eye

photon – light quantum, the smallest possible unit of light; its energy content is greater for high oscillation frequencies and short wavelengths

photopic vision – vision in strong light, using the cones

photoreceptor cells – specialized light-sensitive receptor cells in the retina, two main kinds: rods and cones

pixel – dot-like picture element in electronic image, e.g. in TV screen, computer monitor, camera sensor; commonly, each pixel corresponds to one electronic component (sensor or emitter of light)

primary colours – term with varying definitions; often signifying the starting colours needed for producing other hues with additive (e.g. red, green, blue) or subtractive (e.g. cyan, magenta, yellow) mixing techniques; sometimes used as synonym for *elementary colours*

protanomal – person with an inherited disturbance of red-green colour vision in which the L cones have deviant properties, state: protanomaly protanope – person with an inherited disturbance of red-green colour vision in which the L cones are non-functional; state: protanopia

pupil – round opening for the passage of light at the front of the eye, diameter varying between c. 2 and 8 mm, generally larger in the dark; diameter regulated with radial and circular smooth muscles of the iris

receptive field – for visual nerve cells: region within the visual field at which a change of light gives a change of neuronal activity (increase or decrease)

receptor cells – cells specialized for reacting to a specific kind of physical or chemical influence, different for each kind of receptor cell; connected to the nervous system via particular nerve cells. Visual receptor cells: *rods* and *cones*.

red-blind/red-weak – see *protanope/protanomal*

red-green blindness – term generally used for a group of inherited and common deviations of colour vision, four main varieties (see *deuteranope/deuteranomal*, *protanope/protanomal*); the term is misleading (better alternatives: daltonism or colour uncertainty)

retina – tissue covering the backside inner wall of the eye, contains light-sensitive receptor cells (rods, cones), nerve cells and supporting cells (glia)

retinal – molecule with formula $C_{20}H_{28}O$, resembles vitamin A in chemical composition; retinal and the much larger opsin molecule (a protein) together constitute the visual pigment of a visual receptor cell

rhodopsin – *visual purple*, the visual pigment of rods, coloured purple when in an active state

rod achromat – person lacking functioning cones, vision completely dependent on rods; incapable of discriminating between the hues of chromatic colours; reduced visual function in strong light (e.g. in normal daylight)

rods – the most light-sensitive receptor cells of the retina, used for non-chromatic *scotopic vision* under low-light conditions

S blind – see *tritanope*

S cones – see *cones*

saccades – rapid kind of eye movement, used for directing the eye toward interesting details within the visual field; each saccade is followed by a brief period of fixation, used for collecting and storing the visual image

saturation – relative 'purity' or colourfulness of a chromatic colour (Plate 1.1), maximal for a *monochromatic colour*; saturation decreases with the admixture of achromatic colour (black/grey/white), e.g. by adding light of all other wavelengths

scotopic vision – vision in weak light, using the rods

spectral colours – the monochromatic hues of the sunlight spectrum; according to Newton's classical list: red, orange, yellow, green, blue, indigo, violet (sequence from long- toward shortwave light)

spectrophotometer – instrument for measuring light intensity at different wavelengths

spectrum – sequence of colours seen when light has been separated into its component wavelengths (e.g. using a glass prism); without further specification: spectrum of sunlight

threshold – smallest intensity of stimulation giving a measurable response (e.g. for perception of light or colour, for activation of a nerve cell)

trichromat – person with three types of functional cones, needs three primary colours for admixture of all hues; state: trichromacy

tritanope – person with an inherited disturbance of blue-green colour vision in which the S cones are non-functional; state: tritanopia visual pigment – light-sensitive chemical substance of visual receptor cells, consists of *retinal* and *opsin*; appropriate wavelengths of light will be absorbed, and this alters the structure of retinal and starts a cascade of chemical reactions which cause the receptor cell to change its signalling to nerve cells

visual receptor cells – cones and rods; see *receptor cells*

yellow spot = *macula lutea*, central retinal region covered with the yellow pigment xanthophyll, situated at and around the fovea

Notes

Preface

1. The philosophy of the poem presumably concerns natural philosophy, i.e. natural science.
2. cf. Hardin (1993); Dawkins (1998)

Chapter 1

3. The often stated number of about 16 million colours has the following technical background. The screen of a computer display consists of a large number of very small picture elements (*pixels*), each one of which may be made to emit a light of controlled intensity. Together, all the active pixels form the image seen on the screen. For a colour display, pixels of three different 'primary' colours are intermingled across the screen, each one emitting a light of a different peak-intensity wavelength. The colours seen in the displayed image depend on how the three available pixel-colours are mixed among neighbouring pixels (how this happens will be extensively discussed later on in this book). Usually, each pixel has 256 possible levels of emitted light intensity. Thus, for three different colour-pixels seen together, the number of intensity combinations would be $256 \times 256 \times 256 = 16\ 777\ 216$, which is (mis-)quoted as being equivalent to the same number of colours (see text for further comments).
4. After a period of intense learning and training a maximum of about 50 seems possible (Derefeldt and Berggrund, 1994).
5. Gladstone (1858)
6. See Bellmer (1999)
7. For information on the Munsell system, see Section B.1.1.
8. See also Kay and McDaniel (1997)
9. 'Elementary colours', see Sections 3.2.2 and 4.6.5.
10. e.g. Regier *et al.* (2007)
11. World Color Survey, Berkeley University, USA; the material contains detailed information concerning the basal colour terms and their application in 110 different non-written languages.
12. Gage (1993: p. 79)
13. Gage (1993: pp. 27 and 80)
14. Teller (1998)
15. Villiers and Villiers (1978)
16. Nijboer *et al.* (2007)

17. For readers with a feeling for statistics: hereby the respective standard deviations; each mean $\pm$ SD concerns a group of about 10–15 children. Elementary colours were correctly named for: $67 \pm 2.5\%$ (age 3 years), $75 \pm 3.3\%$ (age 4 years) and $96 \pm 0.4\%$ (age 5 years). For secondary colours the corresponding numbers were: $28 \pm 9.5\%$ (age 3 years), $54 \pm 14.3\%$ (age 4 years), and $95 \pm 0.1\%$ (age 5 years) (data from Nijboer *et al.*, 2007).

18. cf. Frequency of occurrence of different colour terms in fiction (Section 1.9).

19. Dye (2010)

20. Gage (1993: p. 222)

21. *Egyptian blue*, calcium copper silicate ($CaCuSi_4O_{10}$), was called *caeruleum* by the Romans, and is considered to be the first synthetic pigment ever fabricated. Used in Egypt since the 4th Dynasty (2613–2494 BC), popular among the Ancient Greek and Romans, in disuse since 4th century AD. Made by grinding the components, including copper mineral and sand, and then heating the mixture in a furnace.

22. Thompson (1956, p. 102). See also http://en.wikipedia.org/wiki/Minium_%28pigment%29.

23. Valdez and Mehrabian (1994); Suk and Irtel (2010)

24. McManus *et al.* (1981)

25. For references, see McManus *et al.* (1981)

26. Elliot and Niesta (2008); Guéguen (2012)

27 Dreiskaemper *et al.* (2013)

28. Ilie *et al.* (2008)

29. Valdez and Mehrabian (1994). In their summary it is stated that: 'Blue, blue-green, green, red-purple, purple, and purple-blue were the most pleasant hues, whereas yellow and green-yellow were the least pleasant. Green-yellow, blue-green, and green were the most arousing, whereas purple-blue and yellow-red were the least arousing.'

30. Lists with examples of the variability of colour symbolism might be seen in Wikipedia articles concerning various hues (e.g. http://en.wikipedia.org/wiki/Red#Symbolism; http://en.wiki pedia.org/wiki/Green#Symbolism_and_associations; etc.).

31. cf. Treatment with an inert or innocuous substance instead of with a regular medicine; 'placebo' means 'I will please'.

32. e.g. Loftus (1977)

33. See review by Draaisma (2013)

34. The terms 'additive' and 'subtractive' for different types of colour mixing were introduced by the American physicist Ogden Rood in his book *Modern Chromatics, With Applications to Art and Industry* (1879).

35. For explanations and further details, see Section 2.6.2.

36. This concerns mixtures of complementary colours, see Section 2.7.

37. Gage (1993: p. 186)

38. Such devices generally contain a mixture of three kinds of point-like picture elements (pixels), emitting red, green or blue light. In TV screens of various resolution, pixel densities vary between 8.3 and 46 pixels/cm. In other display screens even higher values may occur. See also note 3.

39. Gage (1993: pp. 42 and 64)

40. e.g. Charles Blanc, *Grammaire des Arts du Dessin* (1867); Ogden Rood, *Modern Chromatics, With Applications to Art and Industry* (1879).

41. Gage (1993: p. 11)

42. Gage (1993: p. 15); for an account of modern research into the colours and pigments used in ancient Greek art and examples of coloured reconstructions, see Brinkman *et al.* (2006).
43. Gage 1993: p. 130
44. Gage 1993: p. 71
45. http://sv.wikipedia.org/wiki/Falu_r%C3%B6df%C3%A4rg
46. Locke (1689)
47. Gage (1993: p. 233)
48. Gage (1993: p. 236)
49. Concerning the property of synesthesia, see Section 3.4.1.
50. Some further statistical details concerning Pratt's data: (1) the number of colour terms per 1000 lines of verse were, for the 10 most recent writers 36.1 ± 19.9 (SD), and for the seven earlier ones 13.5 ± 8.5 (difference statistically significant, t test, $p < 0.02$); (2) among all the analysed categories of colour, only the colour blue showed a significantly different frequency of occurrence for later versus earlier writers; mean values were $9.6 \pm 2.3\%$ ($n = 8$) for the later, and $4.7 \pm 2.0\%$ ($n = 5$) for the earlier writers (difference statistically significant, t test, $p < 0.01$). Authors with few colour terms were not included in these latter comparisons.
51. For a detailed recent discussion of philosophical problems concerning colour, see the article on this subject in the web edition of the *Stanford Encyclopedia of Philosophy* (Maund, Barry, 'Color', *The Stanford Encyclopedia of Philosophy* (Winter 2012 Edition), Edward N. Zalta (ed.), URL = http://plato.stanford.edu/archives/win2012/entries/color/).

Chapter 2

52. Gage (1993: p. 162 and Figure 122)
53. Kuehni and Schwarz (2008: p. 45); Gage (1993: p. 166)
54. The term was taken from a Latin word for *appearance/apparition.*
55. Gage (1993: pp. 232 and 168). The 11 steps were: scarlet-purple, red lead, lemon yellow, golden yellow, dark yellow, green, grass green, sea green, blue, indigo, violet.
56. See Section 1.8 for further details concerning Newton's interest in the possible parallels between the tonal scale of music and the colour scale of the sunlight spectrum.
57. Concerns the Archbishop Antonius de Dominis and the philosopher Descartes.
58. Gage (1993: pp. 93–115)
59. Mollon (2003)
60. See Appendix C.1 for examples of calculations.
61. Gage (1993: pp. 42 and 64)
62. For an illustration showing gamuts of different colour spaces, see http://en.wikipedia.org/wiki/Color_space.
63. *Nationalencyklopedin* (1998): article 'Färgtryck' ('Colour printing').
64. Gage (1993); Mollon (2003)
65. Boyle (1664)
66. For more information concerning Hering's system, see Section 3.2.2.
67. See also Section B.1.2 and Figure B.1. In the CIE chart, all colours are complementary that can be joined by a straight line through the central 'white point'. See, for instance, yellow

(a) versus blue (b) in Figure B.1. In colour contexts, the term 'complementarity' was launched in 1793 by Benjamin Thompson (Count Rumford, 1753–1814), who dealt with these phenomena while studying coloured shadows (Section 3.1.3).

68. Nassau (1997)
69. For further details, see Section C.3 and Plate C.1b.
70. In their face, such clocks are labelled with a 'T' near 6 o'clock. Information concerning phosphorescence and light-emitting details in clocks may be found at:

   – http://www.kronometric.org/article/lume/#4.0
   – http://www.uret.se/page_3.html

   (including information concerning the use of tritium (3H))
71. Including dinoflagellates such as *Noctiluca miliaris*, diameter about 2 mm.

Chapter 3

72. See Chevreul (1839)
73. In perception psychology a useful distinction is made between *aperture colour* (as seen in small isolated samples) and *field colour* (as seen in natural scenes with variously coloured objects). Depending on the field/aperture colour context, the precise technical definitions are also somewhat different for the various terms related to brightness (e.g. lightness, value) and colourfulness (chroma, saturation, purity). For further details and definitions, see Hunt and Pointer (2011) and http://www.visualexpert.com/FAQ/Part1/cfaqPart1.html#p1.2
74. See Plate 3.3
75. e.g. Sällström (2001: p. 102)
76. Mollon (2003)
77. Adamson (2001)
78. Land (1997)
79. For recent reports and reviews, see Bartels and Zeki (2000), Shevell and Kingdom (2008), Brainard and Maloney (2011), Foster (2011); see also p. 71 in Stockman and Brainard (2009).
80. Free translation: *Revealing Newton's Colour Theory and an Attempt to Disprove It.*
81. See p. 253–254 in Newton (1730):

   I speak here of colours so far as they arise from light. For they appear sometimes by other causes as when by the power of phantasy we see colours in a dream, or a mad-man sees things before him which are not there; or when we see fire by striking the eye, or see colours like the eye of a peacock's feather, by pressing our eyes in either corner whilst we look the other way. Where these and such like causes interpose not, the colour always answers to the sort or sorts of the rays whereof the light consists, as I have constantly found in whatever phænomena of colours I have hitherto been able to examine.

82. e.g. The Ostwald system (http://www.coloracademy.co.uk/ColorAcademy%202006/subjects/ostwald/ostwald.htm); the Natural Color System (http://www.ncscolour.com).
83. Hurvich (1997)
84. Schefrin and Werner (1990): unique green: mean 507 nm, variation 486–535 nm; unique blue: mean 480 nm, variation 464–495 nm. Cf. Hardin (1993, p. xxiii).

85. Bezold–Brücke hue shift
86. Hurvich (1997: p. 88)
87. Valberg (2005: p. 331)
88. e.g. Yellow was often described as a pale kind of green (Gage 1993: pp. 61, 84 and 119).
89. Moreland and Cruz (1959); Stabell and Stabell (1982)
90. See e.g. *Merriam-Webster's Collegiate Dictionary* (1994)
91. e.g. http://en.wikipedia.org/wiki/Synesthesia
92. Eagleman and Goodale (2009)
93. Gage (1993)
94. http://www.michaelbach.de/ot/col_benham/index.html. An interesting web site, showing examples of many different visual illusions.

Chapter 4

95. Shevell and Kingdom (2008)
96. Hansen and Gegenfurtner (2009)
97. Shevell and Kingdom (2008)
98. In techniques of adaptive optics, computers and adaptive optical equipment are used for decreasing the 'noise' in optical images, thereby causing a significant increase of maximum optical resolution (see Figure 5.1 of Williams and Roorda, 1999). These methods were originally developed for astronomers, but they have also been extremely useful in biomedical research.
99. Average numbers and ratios calculated from the joint results of two papers using the same methods (Roorda and Williams, 1999; Hofer *et al.*, 2005). Only data for male participants ($n = 9$) with normal trichromatic colour vision are included here.
100. Within an image, the analysis of colour requires a comparison between signals from a much larger number of receptors than is needed for the analysis of lightness contrast. To recognize a local contrast of general light intensity, a difference between the activities of two adjacent cones might suffice. To determine a local contrast of colour, the signals must be considered from three kinds of cones on two sides of a potential borderline, i.e. minimally six cones, often more.
101. The central rodless region is about 400–600 μm wide. Its most central portion, lacking S cones, is about 100 μm in diameter, corresponding to a visual angle of about 0.35°. The most central L and M cones are rather thin, which promotes a high central concentration of these visual receptor cells.
102. The great acuity in the centre of the visual field is also highly dependent on neuronal organization: for the central fovea individual bipolar and ganglion cells receive the signals from smaller numbers of receptor cells (down to a 1:1 relation) than what is normally the case more peripherally.
103. This little dimple has a diameter of about 1.5 mm.
104. Loss of image sharpness due to *chromatic aberration* is a common lens problem.
105. One possibility is that the L cones actually have a higher sensitivity than the M cones for light at short wavelengths. In accordance with this hypothesis, measurements have shown a somewhat higher degree of shortwave light absorption for L than for M cones (concerns the

violet part of the spectrum, Plate 4.3b; Dartnall *et al.*, 1983; see also Figure 14 in Gouras 2009). However, this kind of measurement is technically difficult and, hence, somewhat uncertain. Also other possible explanations have been discussed; for instance, it has been suggested that signals from the S cones might perhaps, via retinal neurones, counteract the signals from the M cones and, consequently, increase the relative influence of the L cones at short wavelengths.

106. McIntyre (2002: p. 45)
107. Hansen (2010: p. 89)
108. For a general review, see Geisler (2008).
109. Hansen and Gegenfurtner (2009)
110. The visual pigment of the rods is called *rhodopsin* (also named *visual purple*). It reflects the wavelengths that are not absorbed, i.e. mainly the most longwave plus the most shortwave portions of the spectrum. Together, this combination of long- and shortwave reflected light makes the pigment look purple (when in its active form).
111. See Section C.5 for the intensity of illumination under different conditions, including information about the units used for such measurements. For a camera, a change of light intensity by a factor 1 000 000 000 would correspond to about 30 doublings (or halvings) of exposure time, i.e. in photographic terminology such an alteration of illumination would correspond to about 30 stops.
112. Fain *et al.* (2001); Reuter (2011)
113. A maximum change of pupil diameter by a factor 4 (e.g. from about 2 to 8 mm) would change pupil area by a factor of $4^2 = 16$.
114. Due to desensitizing effects of bleached pigment products (e.g. metarhodopsin III), even a rather limited bleaching of visual pigment might transiently cause a disproportionately large decrease of light sensitivity. Also, in the absence of any regeneration of visual pigment, there is a slow but incomplete recovery of light sensitivity after such bleaching (cf. Figure 9 of Fain *et al.* 2001), due to the gradual disappearance of the desensitizing bleach-products (Fain *et al.* 2001, Reuter 2011). However, a full recovery of light sensitivity requires a regeneration of sufficient amounts of 'active' visual pigment, appropriately provided with 11-cis-retinal.
115. The LGN belongs to a larger neuronal accumulation called *thalamus*.
116. Granit (1970)
117. An interesting statistical example is given on p. 153 and in Figure 6 of Shevell and Kingdom (2008). For estimated degrees of cone activation when viewing a natural scene (e.g. flowers and leaves in a bowl), correlation coefficients were 0.96 for L versus M cones, 0.78 for M versus S, and 0.73 for L versus S. Thus, direct comparisons between coarse activation data for the respective kinds of cone would not have delivered much extra colour information. In contrast, the correlation was only 0.17 for (L + M) versus (L - M), 0.14 for (L + M) versus (S - (L + M)), and −0.16 for (L - M) versus (S - (L + M)). Thus, the calculated cone-activation-*differences* ('colour signals') varied largely independently of each other and independently of the factor (L + M), a measure of general achromatic lightness. The cone difference (L - M) would correspond to hues along the red–green axis, and (S - (L + M)) would correspond to hues along the blue–yellow axis.
118. Brainard *et al.* (2000); Hofer *et al.* (2005)
119. Grimsley (1943)
120. Ikeda and Sagawa (1979)

121. Reviewed in Milner and Goodale (2006). See their Chapter 2 for magno- and parvocellular channels and dorsal and ventral streams; their Chapter 3 for 'blindsight' and associated phenomena.

122. About 10% of the axons from retinal ganglion cells are directed to the brain stem, in particular to the superior colliculus. Part of this visual information reaches the dorsal cortical stream via the pulvinar nucleus of thalamus (see Figure 1.1 and 3.1 in Milner and Goodale, 2006).

123. According to fMRI studies in humans, the V4 complex might be subdivided into two portions with different organization, a posterior V4 and an anterior V4 alpha (Bartels and Zeki, 2000).

124. Zeki (1999); Bartels and Zeki (2000)

125. See Section 2.3.4 in Milner and Goodale (2006).

126. Hubel (1987)

127. The staining concerns the enzyme cytochrome oxidase (CO). In V2, CO-staining reveals a pattern of three types of parallel and alternating stripes: (1) CO-poor (pale) stripes, (2) thick CO-rich stripes, (3) thin CO-rich stripes. The latter type of stripe is rich in colour-selective neurones. General features of the visual pathway and visual cortex are reviewed in Milner and Goodale (2006); CO-staining patterns and the functional organization of V1 and V2 cortices are dealt with in Lu and Roe (2008).

Chapter 5

128. Unfortunately, inherited red-green blindness is sometimes commented upon as if it were a disease (e.g. in the Swedish *Nationalencyklopedin* (1998), article *'Färgblindhet'*). *Encyclopedia Britannica* (1996) uses the more neutral term 'trait'.

129. See p. 29 in Dalton (1798).

130. *Geranium zonule*

131. See p. 264 in Huddart (1777).

132. For instance, Dalton's name is used as a unit for atomic mass:1 dalton = 1 Da = $\sim 1.6605 \times 10^{-27}$ kg

133. Mollon (2003: p. 10)

134. Rayleigh's measuring device was a predecessor of the modern anomaloscope (Rayleigh, 1890). Investigations by the Dutch ophthalmologist F.C. Donders were also important for clarifying the properties of anomalous trichromats (Mollon, 2003). For further information concerning the anomaloscope, see Section A.4 and Plate A.2.

135. DeMarco *et al.* (1992)

136. e.g. Failing the Ishihara test but not the D15; see Plates 5.1 and A.1, Sections A.1 and A.2.

137. Sharpe *et al.* (1999: pp. 38–39)

138. Sharpe *et al.* (1999: p. 28)

139. Thus >525 versus <470 nm.

140. McIntyre (2002: p. 43)

141. Various web sites, including www.vischeck.com, demonstrate how colours might be perceived by various categories of colour-blind people as compared to normal trichromats. See text for comments on the uncertainties involved in such endeavours.

142. See e.g. Graham and Hsia (1959)

143. Chapanis (1944); Breton and Tansley (1985); Derefeldt and Berggrund (1994: p. 100)

144. Chapanis (1944)

145. Hansen (2010: p. 66); see also Section 5.5.1

146. Large-field trichromacy might, for instance, be studied by comparing the colour discrimination for a field with a visual angle of $>8°$ to that for smaller fields (e.g. $2°$). For a review and discussion of possible causes for the effects of field size on colour discrimination, see Sharpe *et al.* (1999: p. 34). See also Smith and Pokorny (1977); Pokorny and Smith (1982).

147. e.g. Various electrical appliances, including some DVD recorders and digital alarm clocks.

148. $\sim$498 nm versus $\sim$492 nm; Sharpe *et al.* (1999: p. 29); cf. Hsia and Graham (1997: p. 203).

149. Morgan *et al.* (1992)

150. Deviant Rayleigh match, see Appendix A.4, Plate A.2.

151. Bosten *et al.* (2005)

152. Melin *et al.* (2007)

153. Verhulst and Maes (1998)

154. Simunovic *et al.* (2001)

155. McIntyre (2002: p. 97)

156. McIntyre (2002: p. 140)

157. Steward and Cole (1989)

158. There are many websites giving recommendations concerning suitable and non-suitable colour combinations for sites intended to be easily (or at all) readable for the whole population, including those with red-green blind eyes. A few examples of such sites:

    - http://msdn.microsoft.com/en-us/library/bb263953(v=vs.85).aspx
    - http://webdesign.about.com/od/accessibility/a/aa062804.htm
    - http://jfly.iam.u-tokyo.ac.jp/color/

159. e.g. A pseudoisochromatic test like Ishihara plus, for the colour-blind children, a Farnsworth D15 test; see Appendix A.

160. Verriest *et al.* (1980); McIntyre (2002: p. 133)

161. McIntyre (2002: p. 134)

162. In Japan, applicants for a driving licence have to pass a practical test concerning colour vision. For a foreigner, it may, in this context, seem confusing that the colour of the green Japanese traffic lights is named using a term that often means 'blue'. Unfortunately, the traffic lights remain green and the unexpected colour naming has cultural–historical–linguistic reasons; for the complex relationships between language and colour, see Section 1.2.

163. Nilsson (2004: p. 49)

164. Hansen (2010: p. 117)

165. Steward and Cole (1989)

166. Translated from Swedish by the author; sentence cited from p. 84 of Holmgren (1877).

167. cf. http://sv.wikipedia.org/wiki/Lagerlundaolyckan

168. Holmgren's test, also known as the *Holmgren–Thompson test*, was one of the first tests for colour blindness to be commercially marketed (until recently still available). It consisted of a series of wool skeins in different colours. For further information, see Introduction of Appendix A and Holmgren (1877). For a description of the test and an example of how it was used, see http://www.psych.utoronto.ca/museum/holmgren.htm.

169. For both the Apple and Android operating systems, many such applications are available and easily found on the internet. The apps may provide an RGB code as well as a name for the (mean) color within a restricted region of the camera view. Some apps may even speak out the colour name, making the information useful also for the totally blind (e.g. when sorting/selecting clothes). For identifying code colours of point-like sources of light (e.g. LEDs of a battery charger) such apps are, however, less useful than optical filters.
170. e.g. Seebeck (1837); Maxwell (1855); review Sharpe and Jägle (2001).
171. Kernell (1974); Independently of each other, similar methods have been developed and tested several times by different people. For a commercially marketed variety of handheld filters for the red-green blind, see www.Seekey.se.
172. Kernel (1974)
173. e.g. The encoding colours of electrical resistors (Kernell, 1974).
174. e.g. Maxwell (1855)
175. McIntyre (2002: p. 129)
176. Jacobs *et al.* (2007); Jacobs and Nathans (2009)
177. Mancuso *et al.* (2009)

Chapter 6

178. Sharpe *et al.* (1999: p. 45)
179. Hansen (2010: p. 41)
180. Two out of several web sites concerning rod achromatopsia:

  – for 'The achromatopsia network' – http://www.achromat.info/
  – for 'The achromatopsia group' – http://www.achromatopsia.org/.

181. References and further information: Sharpe *et al.* (1999: p. 50).
182. Cole *et al.* (2003)
183. Kentridge *et al.* (2004)
184. e.g. Geschwind and Fusillo (1997)
185. McIntyre (2002: p. 55); Hansen (2010: p. 46)
186. Hansen (2010: pp. 48 and 120–124)
187. McIntyre (2002: pp. 56 and 116); Heaton *et al.* (1958)
188. Sharpe *et al.* (1999: p. 14)
189. Jameson *et al.* (2001)
190. Jordan and Mollon (1993)
191. Jordan *et al.* (2010)

Chapter 7

192. Land and Nilsson (2002: p. 4)
193. Onishi *et al.* (1999)

194. Jacobs and Williams (2001)
195. Peichl *et al.* (2001)
196. Glösmann *et al.* (2008)
197. Tritsch (1993)
198. Arrese *et al.* (2002)
199. *Platypus* has only two kinds of cones, one of which contains a phylogenetically very old visual pigment for short wavelengths of light (Davies *et al.* 2007).
200. Jacobs and Deegan (1993)
201. Collin and Trezise (2004)
202. e.g. Red droplets in cones for long wavelengths; cf. colour analysis in sensors of digital cameras (Section D.1).
203. *Bubo virginianus*, Jacobs *et al.* (1987)
204. Land and Nilsson (2002: p. 65)
205. Land and Nilsson (2002: p. 98)
206. Briscoe and Chittka (2001)
207. Land and Nilsson (2002: p. 130)
208. Cronin *et al.* (2000); Land and Nilsson (2002: p. 195). For recent studies of colour vision and behaviour in mantis shrimps, see Thoen *et al.* (2014).
209. Thoen et al. (2014)
210. Rayleigh scattering, Section 2.8.2.2
211. Land and Nilsson (2002: p. 29)

Appendix A

212. For further details, illustrations and references, see Birch (2003); McIntyre (2002); Hansen (2010).
213. Hansen (2010: p. 23)
214. For further information, see p. 59 in Birch (2003).
215. Birch (2003)
216. The firm Richmond Products offers a great variety of testing material concerning colour vision and other visual functions, including a magnet-operated version of the D15 test. Their site: http://www.richmondproducts.com.
217. See Vingrys and Cole (1986: p. 371).
218. In this context, it is important to remember that the colour 'green' covers a wide region in the spectrum and that different varieties (wavelengths) of 'green' will give quite different results when additively mixed with red. As one might see in the CIE chromaticity chart (Plate B.2; Figure B.1) red and 546 nm green lie along the same straight line with yellow in between. Consequently, a mixture of red and 546 nm green may become yellow but never white; this green and red are not complementary colours. A straight line from red through the white point will reach the periphery of the CIE curve near 495 nm green. This is the green which is complementary to red; 495 nm green plus red may become white but never yellow.
219. Birch (2003: p. 24).
220. Sharpe *et al.* (1999: p. 29)

221. Provided that the traffic lights have properties in accordance with the CIE recommendation *green class A*, which is often the case. For an illustration of how this signal colour is situated in the CIE chromaticity diagram, see Figure 3 of Cole (2004).
222. Hunt *et al.* (1995)
223. Current capabilities of genetical tests for various inborn varieties of deviant colour vision are described at the web site of the American National Institutes of Health (NIH): http://www .ncbi.nlm.nih.gov/books/NBK1301/.

Appendix B

224. Kuehni and Schwarz (2008)
225. See p. 78–80 in Kuehni and Schwarz (2008)
226. Web page for the company selling products associated with the Munsell system: http://www .xrite.com/top_munsell.aspx. See also www.MunsellStore.com for publications. For infor- mation about the *Munsell Book of Color*, see http://www.xrite.com/product_overview.aspx? ID=864.
227. *Ibid.*
228. *Ibid.*
229. Web site for the firm selling the NCS system: http://ncscolour.com/. At this site, the NCS Navigator tool is very useful for investigating the properties and organization of the NCS system.
230. Web site: www.pantone.com.
231. Colour specifications for the Swedish flag: http://flagspot.net/flags/se_fact.html#d.
232. Cf. p. 10.10 and Figure 5 in Brainard and Stockman (2009). Models based on the original CIE tricolour-matching measurements had fictive colour sensors with partly 'negative' reactions to some wavelengths of light, meaning that in order to get a good colour match it was sometimes necessary to add some coloured mixing light to the test light. This aspect of the model was not physiologically attractive, and the model was therefore adapted such as to get only positive characteristics; this was done using a mathematical transformation of the RGB-light sources.
233. Concerning the effects of shortwave light on L versus M cones, see also Section 4.3.3 and note 105.
234. See e.g. Kuehni and Schwarz (2008); Brainard and Stockman (2009); Hunt and Pointer (2011).
235. See: http://en.wikipedia.org/wiki/Lab_color_space.
236. For further information and calculation methods, see http://en.wikipedia.org/wiki/ LMS_color_space.

Appendix C

237. Valberg (2005, p. 169)
238. For further details and comments, see e.g. www.scopecalc.com.
239. Valberg (2005, p. 105)

Appendix D

240. For digital cameras, one of the widely used versions of the RGB system is called sRGB. There are also other versions, including the Adobe-RGB (marketed by the Adobe company).

References

As a support for further studies, I will first give a list with a commented selection of useful and important reviews. In the running text and in illustration legends, I also cite some more specialized articles and books (often concerning items not treated in the cited reviews; references often given in associated notes). All publications cited below are included in the Cited References list. In addition, I have used information gathered from general encyclopedias (e.g. *Encyclopedia Britannica*; the Swedish *Nationalencyklopedin*) and from various web-versions of Wikipedia (as far as possible with data confirmed also from other sources).

Selected literature (with comments)

Chapter 1. Colour vision in everyday life

- **Byrne and Hilbert** (eds) (1997a): *Readings on Color. 1. The Philosophy of Color.* A general survey of how present-day philosophers think about the concept and phenomenon of colour.
- **Finlay** (2003): *Kleur, een reis door de geschiedenis* (original English title: *Colour. Travels Through the Paintbox*). Entertaining description of human efforts to find and produce good colour pigments.
- **Gage** (1993): *Colour and Culture. Practice and Meaning from Antiquity to Abstraction.* Contains a huge amount of (sometimes curious) cultural/historical detailed information concerning colour; important source of information for Chapters 1, 2 and 3. The book is beautifully illustrated, it contains 2408 literature references, and it is (according to the author himself) the result of more than 30 years of work.
- **Hardin** (1993): *Color for Philosophers. Unweaving the Rainbow.* A book of interest for Chapter 1, as well as for several others. Written by an American philosopher, it deals with the borderline region of philosophy and psychology/physiology, contains much empirical scientific information.
- **Kuehni and Schwarz** (2008): *Color Ordered. A Survey of Color Order Systems from Antiquity to the Present.* Important source of information for Chapters 1, 2, 3 and Appendix B. Contains an impressive amount of illustrations, descriptions and physiological comments concerning different kinds of colour systems, from ancient times to the present.
- **Nilson** (2004): *KG Nilsons färglära.* A Swedish book, written by an artist, includes interesting general information concerning colours, colour pigments, and their practical use.

Chapters 2–3. Colour signals and visual fields

Historical background (for the associated physiology, see references for Chapter 4)

- Boyle (1664): *Experiments and Considerations Touching Colours*. An example of pre-Newtonian thoughts and observations concerning colour, written by a famous physicist. Book available at Gutenberg.org, Ebook #14504. Paradoxically, such ancient classics have become easily available via the internet, thanks to the modern development of electronic media. This is also true for:
- Newton (1730): *Opticks: Or a Treatise of the Reflections, Refractions, Inflections and Colours of Light*. A classical work that is still readable; this is the last edition that was edited by Newton himself (Gutenberg.org, EBook #33504); the first edition was printed in 1704.
- Mollon (2003): *The Origins of Modern Color Science*. Very useful review concerning the development of colour science (available on the internet).

Chapter 4. Our biological hardware: eye and brain

- Hubel (1987): *Eye, Brain and Vision*. General summary written by prominent visual neurophysiologist concerning the neuronal processing of visual information. http://hubel.med.harvard.edu/
- Kolb, Fernandez and Nelson (2015): *Webvision: The Organization of the Retina and Visual System* (http://www.webvision.med.utah.edu/). A general and very detailed summary, freely available on the internet. Includes a chapter on colour vision written by Gouras, a research scientist in this field.
- Valberg (2005): *Light Vision Color*. A very useful and readable general review of modern visual science, including colour vision.

Specifically concerning colour vision

The following three volumes give a good survey of modern research concerning colour vision and its deviations; each book contains several articles written by different scientists.

- Byrne and Hilbert (eds) (1997b): *Readings on Color. 2. The Science of Color*.
- Gegenfurtner and Sharpe (eds) (1999): *Color Vision. From Genes to Perception*.
- Mollon, Pokorny and Knoblauch (eds) (2003): *Normal and Defective Colour Vision*.

Two recently published reviews by prominent investigators of the visual system (available via www.cvrl.org –> Publications –> Stockman):

- Brainard and Stockman (2009): *Colorimetry*.
- Stockman and Brainard (2009): *Color Vision Mechanisms*.

Special subjects

- Findlay and Gilchrist (2003): *Active Vision. The Psychology of Looking and Seeing*. Includes an analysis of saccadic eye movements and their importance.
- Milner and Goodale (2006): *The Visual Brain in Action*. A detailed survey of the organization of conscious and non-conscious visual processing in the brain.
- Zeki (1999): *Inner Vision. An Exploration of Art and the Brain*. An internationally well-known neurophysiologist writes on the fascinating border region between brain physiology and perception psychology, focussing on pictorial art. The brain's processing of colour is one of the major questions dealt with in this book.

Chapters 5–6. Deviant colour vision

Historical background

- **Dalton** (1798): *Extraordinary Facts Relating to the Vision of Colours.* This is the first detailed scientific description of red-green blindness. John Dalton analyses his own colour vision in relation to that of others. His paper was published on pages 28–45 of Volume 5 (Part 1) of *Memoirs of the Literary and Philosophical Society of Manchester (1798)*, which thanks to Google Books, is freely available via the internet.
- **Holmgren** (2010[1877]): *Om färgblindheten i dess förhållande till jernvägstrafiken och sjöväsendet.* This is a recent re-publication of an early Swedish scientific report concerning colour blindness in relation to the coloured signals used for railway traffic and at sea.

Present-day accounts

Current investigations concerning colour blindness are thoroughly treated in the reviews mentioned for Chapter 4. In addition, two interesting books are listed below, both specifically dealing with colour blindness and both written for a general readership (although the first one is in Norwegian).

- **Hansen** (2010): *Färgeblindhet.* Written by a Norwegian ophthalmologist with clinical experience concerning colour blinds and their problems.
- **McIntyre** (2002): *Colour Blindness. Causes and Effects.* Written by a British physicist, including comments on his own colour blindness.

Chapter 7. Colour vision in different species of animal

- **Land and Nilsson** (2002): *Animal Eyes.* A very interesting general review of vision in various animal species, vertebrates as well as invertebrates. Includes some information on colour vision.

Appendix A. Tests concerning colour vision

- **Birch** (2003): *Diagnosis of Defective Colour Vision.* A thorough and very useful review, packed with interesting quantitative information concerning colour vision in normal trichromats and in various kinds of colour blindness.

Appendices B, C and D. Colour systems and measurements; light and illumination; digital cameras

For these subjects, the books by Valberg (2005), Kuehni and Schwarz (2008), Hunt and Pointer (2011) and Westland (2012) are very useful.

Cited references

Adamson, J. C. (2001). Edwin Land's interesting accident: a 2-color system. Available at: www.greatreality.com/Color2Color.htm (accessed 7 May 2015).

Arrese, C. A., Hart, N. S., Thomas, N., Beazley, L. D. and Shand, J. (2002). Trichromacy in Australian marsupials. *Current Biology*, **12**, 657–660.

Bartels, A. and Zeki, S. (2000). The architecture of the colour centre in the human visual brain: new results and a review. *European Journal of Neuroscience*, **12**, 172–193.

Bellmer, E. H. (1999). The statesman and the ophthalmologist: Gladstone and Magnus on the evolution of human colour vision, one small episode of the nineteenth-century Darwinian debate. *Annals of Science*, **56**, 25–45.

Berlin, B. and Kay, P. (1969). *Basic Color Terms: Their Universality and Evolution*. Berkley and Los Angeles CA: University of California Press.

Birch, J. (2003). *Diagnosis of Defective Colour Vision*. Edinburgh, UK: Butterworth-Heinemann.

Björk, S. (1954). Ut i det blå, In *Karl XII:s stövlar*. Stockholm, Sweden: Bonnier. pp. 159–175.

Blanc, C. (1867). *Grammaire des Arts du Dessin*. Paris: Ve J. Renouard.

Bosten, J. M., Robinson, J. D., Jordan, G. and Mollon, J. D. (2005). Multidimensional scaling reveals a color dimension unique to 'color-deficient' observers. *Current Biology*, **15**, R950–R952.

Boyle, R. (1664). *Experiments and Considerations Touching Colours*. London, UK: Herringman. (Available as electronic reprint: project Gutenberg, EBook #14504.)

Brainard, D. H. and Maloney, L. T. (2011). Surface color perception and equivalent illumination models. *Journal of Vision*, **11**(5), 1–18.

Brainard, D. H. and Stockman, A. (2009). Colorimetry. In Bass, M., DeCusatis, C., Enoch, J., *et al.* (eds), *The Optical Society of America Handbook of Optics, 3rd edition, Volume III: Vision and Vision Optics*. New York: McGraw Hill, pp. 10.1–10.56.

Brainard, D. H., Roorda, A., Yamauchi, Y. *et al.* (2000). Functional consequences of the relative numbers of L and M cones. *Journal of the Optical Society of America A. Optics, image science, and vision*, **17**, 607–614.

Breton, M. E. and Tansley, B. W .(1985). Improved color test results with large-field viewing in dichromats. *Archives of Ophthalmology*, **103**, 1490–1495.

Brinkman, V., Brijder, H. *et al.* (2006). *Kleur! bij Grieken en Etrusken*. Zwolle, The Netherlands: Waanders Uitgevers.

Briscoe, A. D. and Chittka, L. (2001). The evolution of color vision in insects. *Annual Review of Entomology*, **46**, 471–510.

Byrne, A. and Hilbert, D. R. (eds) (1997a). *Readings on Color. 1. The Philosophy of Color.* Cambridge MA: MIT Press.

Byrne, A. and Hilbert, D. R. (eds) (1997b). *Readings on Color. 2. The Science of Color.* Cambridge MA: MIT Press.

Carroll, J., Murphy, C. J., Neitz, M., Ver Hoeve, J. N. and Neitz, J. (2001). Photopigment basis for dichromatic color vision in the horse. *Journal of Vision*, **1**, 80–87.

Chapanis, A. (1944). Spectral saturation and its relation to color-vision defects. *Journal of Experimental Psychology*, **34**, 24–44.

Chen, D.-M. and Goldsmith, T. H. (1986). Four spectral classes of cone in the retinas of birds. *Journal of Comparative Physiology A*, **159**, 473–479.

Chevreul, M.-E. (1839). *De la loi du contraste simultané des couleurs*. Paris, France: Pitois-Levrault.

Cole, B. L. (2004). The handicap of abnormal colour vision. *Clinical and Experimental Optometry*, **87**, 258–275.

Cole, G. G., Heywood, C., Kentridge, R., Fairholm, I. and Cowey, A. (2003). Attentional capture by colour and motion in cerebral achromatopsia. *Neuropsychologia*, **41**, 1837–1846.

Collin, S. P. and Trezise, A. E. (2004). The origins of colour vision in vertebrates. *Clinical and Experimental Optometry*, **87**, 217–223.

Crognale, M. A., Levenson, D. H., Ponganis, P. J., Deegan, J. F. I. and Jacobs, G. H. (1998). Cone spectral sensitivity in the harbor seal (*Phoca vitulina*) and implications for color vision. *Canadian Journal of Zoology*, **76**, 2114–2118.

Cronin, T. W., Marshall, N. J. and Caldwell, R. L. (2000). Spectral tuning and the visual ecology of mantis shrimps. *Philosophical Transactions of the Royal Society London B*, **355**, 1263–1267.

Dalton, J. (1798). Extraordinary facts relating to the vision of colours. *Memoirs of the Literary and Philosophical Society of Manchester*, **5**, 28–45.

Dartnall, H. J. A., Bowmaker, J. K. and Mollon, J. D. (1983). Human visual pigments: microspectrophotometric results from the eyes of seven persons. *Proceedings of the Royal Society London B*, **220**, 115–130.

Davies, W. L., Carvalho, L. S., Cowing, J. A., *et al.* (2007). Visual pigments of the platypus: a novel route to mammalian colour vision. *Current Biology*, **17**, R161–R163.

Dawkins, R. (1998). *Unweaving the Rainbow: Science, Delusion and the Appetite for Wonder.* Boston MA: Houghton Mifflin.

DeMarco, P., Pokorny, J. and Smith, V. C. (1992). Full-spectrum cone sensitivity functions for X-chromosome-linked anomalous trichromats. *Journal of the Optical Society of America A*, **9**, 1465–1476.

Derefeldt, G. and Berggrund, U. (1994). *Färg som informationsbärare.* Stockholm, Sweden: Försvarets Forskningsanstalt.

Draaisma, D. (2013). *De dromenwever.* Groningen, The Netherlands: Historische Uitgeverij.

Dreiskaemper, D., Strauss, B., Hagemann, N. and Büsch, D. (2013). Influence of red jersey color on physical parameters in combat sports. *Journal of Sport and Exercise Psychology*, **35**, 44–49.

Dye, M. (2010). Why Johnny can't name his colors. *Scientific American.* Available at: www.scientificamerican.com/article.cfm?id=why-johnny-name-colorsandpage=2 (accessed 7 May 2015).

Eagleman, D. M. and Goodale, M. A. (2009). Why color synesthesia involves more than color. *Trends in Cognitive Sciences*, **13**, 288–292.

Elliot, A. J. and Niesta, D. (2008). Romantic red: red enhances men's attraction to women. *Journal of Personality and Social Psychology*, **95**, 1150–1164.

Fain, G. L., Matthews, H. R., Cornwall, M. C. and Koutalos, Y. (2001). Adaptation in vertebrate photoreceptors. *Physiological Reviews*, **81**, 117–151.

Findlay, J. M. and Gilchrist, I. D. (2003). *Active Vision. The Psychology of Looking and Seeing.* Oxford, UK: Oxford University Press.

Finlay, V. (2003). *Kleur, een reis door de geschiedenis [Colour. Travels through the Paintbox].* Amsterdam, The Netherlands: Ambo/Anthos.

Foster, D. H. (2011). Color constancy. *Vision Research*, **51**, 674–700.

Gage, J. (1993). *Colour and Culture. Practice and Meaning from Antiquity to Abstraction.* London, UK: Thames and Hudson.

Gegenfurtner, K. R. and Sharpe, L. T. (eds) (1999). *Color Vision. From Genes to Perception.* Cambridge, UK: Cambridge University Press.

Geisler, W. S. (2008). Visual perception and the statistical properties of natural scenes. *Annual Review of Psychology*, **59**, 167–192.

Geschwind, N. and Fusillo, M. (1997). Color-naming defects in association with alexia. In Byrne, A. and Hilbert, D. R. (eds), *Readings on Color. 2. The Science of Color.* Cambridge MA: MIT Press, pp. 261–275.

Gladstone, W. E. (1858). *Studies on Homer and the Homeric Age*. Vol III. Oxford, UK: Oxford University Press.

Glösmann, M., Steiner, M., Peichl, L. and Ahnelt, P. K. (2008). Cone photoreceptors and potential UV vision in a subterranean insectivore, the European mole. *Journal of Vision*, 8, 1–12.

Goethe, J. W. (1810). *Zur Farbenlehre*. Tübingen, Germany: J. G. Cotta'schen Buchhandlung.

Gouras, P. (2009). Color vision. In Kolb, H., Fernandez, E. and Nelson, R. (eds), *Webvision: The Organization of the Retina and Visual System*. Salt Lake City UT: John Morgan Eye Center, University of Utah. Available at: www.webvision.med.utah.edu/Color.html (accessed 7 May 2015).

Graham, C. H. and Hsia, Y. (1959). Studies of color blindness: a unilaterally dichromatic subject. *Proceedings of the National Academy of Sciences*, 45, 96–99.

Granit, R. (1970). The development of retinal neurophysiology. *Nobel Lectures, Physiology or Medicine 1963–1970*. Amsterdam, The Netherlands: Elsevier.

Grimsley, G. (1943). A study of individual differences in binocular color fusion. *Journal of Experimental Psychology*, 32, 82–87.

Guéguen, N. (2012). Color and women attractiveness: when red clothed women are perceived to have more intense sexual intent. *Journal of Social Psychology*, 152, 261–265.

Hansen, E. (2010). *Färgeblindhet*. Oslo, Norway: Gyldendal Akademisk.

Hansen, T. and Gegenfurtner, K. R. (2009). Independence of color and luminance edges in natural scenes. *Visual Neuroscience*, 26, 35–49.

Hardin, C. L. (1993). *Color for Philosophers. Unweaving the Rainbow*. Indianapolis IN: Hackett.

Hart, N. S., Partridge, J. C. and Cuthill, I. C. (1999). Visual pigments, cone oil droplets, ocular media and predicted spectral sensitivity in the domestic turkey (*Meleagris gallopavo*). *Vision Research*, 39, 3321–3328.

Heaton, J. M., McCormick, A. J. A. and Freeman, A. G. (1958). Tobacco amblyopia: a clinical manifestation of vitamin B12 deficiency. *Lancet* 2 (7041), 286–290.

Hofer, H., Carroll, J., Neitz, J., Neitz, M. and Williams, D. R. (2005). Organization of the human trichromatic cone mosaic. *Journal of Neuroscience*, 25, 9669–9679.

Holmgren, F. (1877). *Om färgblindheten i dess förhållande till jernvägstrafiken och sjöväsendet*. Uppsala, Sweden: Ed. Berlings boktryckeri.

Hsia, Y. and Graham, C. H. (1997). Color blindness. In Byrne, A. and Hilbert, D. R. (eds), *Readings on Color. 2. The Science of Color*. Cambridge MA: MIT Press, pp. 201–229.

Hubel, D. H. (1987). *Eye, Brain and Vision*. Scientific American Library. Available at: http://hubel .med.harvard.edu/.

Huddart, J. (1777). An account of persons who could not distinguish colours. *Philosophical Transactions of the Royal Society of London*, 67, 260–265.

Hunt, D. M., Dulai, K. S., Bowmaker, J. K. and Mollon, J. D. (1995). The chemistry of John Dalton's color blindness. *Science*, 267, 984–988.

Hunt, R. W. G. and Pointer, M. R. (2011). *Measuring Colour*. Chichester, UK: Wiley.

Hurvich, L. M. (1997). Chromatic and achromatic response functions. In Byrne, A. and Hilbert, D. R. (eds), *Readings on Color. 2. The Science of Color*. Cambridge MA: MIT Press, pp. 67–91.

Ikeda, M. and Sagawa, K. (1979). Binocular color fusion limit. *Journal of the Optical Society of America*, 69, 316–320.

Ilie, A., Ioan, S., Zagrean, L. and Moldovan, M. (2008). Better to be red than blue in virtual competition. *CyberPsychology and Behavior*, 11, 375–377.

Jacobs, G. H. (1996). Primate photopigments and primate color vision. *Proceedings of the National Academy of Sciences*, **93**, 577–581.

Jacobs, G. H. and Deegan, J. F. (1993). Photopigments underlying color vision in ringtail lemurs (*Lemur catta*) and brown lemurs (*Eulemur fulvus*). *American Journal of Primatology*, **30**, 243–256.

Jacobs, G. H. and Nathans, J. (2009). The evolution of primate color vision. *Scientific American*, **300**, 56–63.

Jacobs, G. H. and Williams, G. A. (2001). The prevalence of defective color vision in Old World monkeys and apes. *Color Research and Application*, **26**, S123–S127.

Jacobs, G. H., Crognale, M. and Fenwick, J. (1987). Cone pigment of the great horned owl. *The Condor*, **89**, 434–436.

Jacobs, G. H., Deegan, J. F. I. and Neitz, J. (1998). Photopigment basis for dichromatic color vision in cows, goats and sheep. *Visual Neuroscience*, **15**, 581–584.

Jacobs, G. H., Fenwick, J. A. and Williams, G. A. (2001). Cone-based vision of rats for ultraviolet and visible lights. *Journal of Experimental Biology*, **204**, 2439–2446.

Jacobs, G. H., Williams, G. A., Cahill, H. and Nathans, J. (2007). Emergence of novel color vision in mice engineered to express a human cone photopigment. *Science*, **315**, 1723–1725.

Jameson, K. A., Highnote, S. M. and Wasserman, L. M. (2001). Richer color experience in observers with multiple photopigment opsin genes. *Psychonomic Bulletin and Review*, **8**, 244–261.

Jordan, G. and Mollon, J. D. (1993). A study of women heterozygous for colour deficiencies. *Vision Research*, **33**, 1495–1508.

Jordan, G., Deeb, S. S., Bosten, J. M. and Mollon, J. D. (2010). The dimensionality of color vision in carriers of anomalous trichromacy. *Journal of Vision*, **10**(8):12, 1–19.

Kay, P. and McDaniel, C. K. (1997). The linguistic significance of the meanings of basic color terms. In Byrne, A. and Hilbert, D. R. (eds), *Readings on Color. 2. The Science of Color*. Cambridge MA: MIT Press, pp. 399–441.

Kentridge, R. W., Heywood, C. A. and Cowey, A. (2004). Chromatic edges, surfaces and constancies in cerebral achromatopsia. *Neuropsychologia*, **42**, 821–830.

Kernell, D. (1974). A simple and inexpensive method for helping 'red-green blind' subjects to identify colours. *Journal of Physiology*, **241**, 73–74P.

Kolb, H., Fernandez, E. and Nelson, R. (2015). *Webvision: The Organization of the Retina and Visual System*. John Morgan Eye Center, University of Utah, Salt Lake City UT. Available at: www.webvision.med.utah.edu/ (accessed 7 May 2015).

Kuehni, R. G. and Schwarz, A. (2008). *Color Ordered. A Survey of Color Order Systems from Antiquity to the Present*. Oxford, UK: Oxford University Press.

Land, E. H. (1997). Recent advances in Retinex theory. In Byrne, A. and Hilbert, D. R. (eds), *Readings on Color. 2. The Science of Color*. Cambridge MA: MIT Press, pp. 143–159.

Land, M. F. and Nilsson, D.-E. (2002). *Animal Eyes*. Oxford, UK: Oxford University Press.

Locke, J. (1689). An essay concerning human understanding. Available at: http://ebooks.adelaide .edu.au/l/locke/john/l81u/B3.4.html, accessed 7 May 2015.

Loftus, E. (1977). Shifting human color memory. *Memory and Cognition*, **5**, 696–699.

Lu, H. D. and Roe, A. W. (2008). Functional organization of color domains in V1 and V2 of macaque monkey revealed by optical imaging. *Cerebral Cortex*, **18**, 516–533.

Mancuso, K., Hauswirth, W. W., Quihong, L., *et al.* (2009). Gene therapy for red-green colour blindness in adult primates. *Nature*, **461**, 784–787.

Maxwell, J. C. (1855). Experiments on colour, as perceived by the eye, with remarks on colour-blindness. *Transactions of the Royal Society of Edinburgh*, **21**, 275–289 [pp. 126–154 in Maxwell 1890].

Maxwell, J. C. (1860). On the theory of compound colours, and the relations of the colours of the spectrum. *Philosophical Transactions* [pp. 410–444 in Maxwell 1890].

Maxwell, J. C. (1890). *The Scientific Papers of James Clerk Maxwell.* Cambridge, UK: Cambridge University Press.

McIntyre, D. (2002). *Colour Blindness. Causes and Effects.*, Chester, UK: Dalton.

McManus, I. C. (1983). Basic colour terms in literature. *Language and Speech*, **26**, 247–252.

McManus, I. C., Jones, A. L. and Cottrell, J. (1981). The aesthetics of colour. *Perception*, 10, 651–666.

Melin, A. D., Fedigan, L. M., Hiramatsu, C., Sendall, C. L. and Kawamura, S. (2007). Effects of colour vision phenotype on insect capture by a free-ranging population of white-faced capuchins, *Cebus capucinus. Animal Behaviour*, **73**, 205–214.

Merriam-Webster (1994). *Merriam-Webster's Collegiate Dictionary.* Springfield MA: Merriam-Webster, Inc.

Milner, R. D. and Goodale, M. A. (2006). *The Visual Brain in Action.* 2nd edn. Oxford, UK: Oxford University Press.

Mollon, J. D. (2003). The origins of modern color science. In Shevell, S. K. (ed.), *The Science of Color.* Amsterdam, The Netherlands: Elsevier, pp. 1–39.

Mollon, J. D., Pokorny, J. and Knoblauch, K. (eds) (2003). *Normal and Defective Colour Vision.* Oxford, UK: Oxford University Press.

Moreland, J. D. and Cruz, A. (1959). Colour perception with the peripheral retina. *Journal of Modern Optics*, **6**, 117–151.

Morgan, M. J., Adam, A. and Mollon, J. D. (1992). Dichromats detect colour-camouflaged objects that are not detected by trichromats. *Proceedings of the Royal Society London B*, **248**, 291–295.

Nassau, K. (1997). The causes of color. In Byrne, A. and Hilbert, D. R. (eds), *Readings on Color. 2. The Science of Color.* Cambridge MA: MIT Press, pp. 3–29.

Neitz, J., Geist, T. and Jacobs, G. H. (1989). Color vision in the dog. *Visual Neuroscience*, **3**, 119–125.

Newton, I. (1704). *Opticks: or a Treatise of the Reflexions, Refractions, Inflexions and Colours of Light.* London: Smith and Walford.

Newton, I. (1730). *Opticks: or a Treatise of the Reflections, Refractions, Inflections and Colours of Light.* London: William Innys.

Nijboer, T. C. W., Smagt, M. J. vd, Bullens, J., Haan, E. H. F. d and Zandvoort, M. J. E. v (2007). Distinct trajectories for acquisition of colour terms and object-colour-associations. In Nijboer, T. C. W. (ed.), *Neuropsychology of Colour Vision. Studies in Patients with Acquired Brain Damage, Healthy Participants, and Cases with Developmental Disorders.* Utrecht University, Utrecht, The Netherlands, pp. 89–104.

Nilson, K. G. (2004). *KG Nilsons färglära.* Stockholm, Sweden: Carlssons.

Nordby, K. (1996). Vision in a complete achromat: a personal account. Available at: http://consc.net/misc/achromat.html (accessed 7 May 2015).

Onishi, A., Koike, S., Ida, M., *et al.* (1999). Dichromatism in macaque monkeys. *Nature*, **402**, 139–140.

Palacios, A. G., Varela, F. J., Srivastava, R. and Goldsmith, T. H. (1998). Spectral sensitivity of cones in the goldfish, *Carassius auratus. Vision Research*, **38**, 2135–2146.

Peichl, L., Behrmann, G. and Kröger, R. H. H. (2001). For whales and seals the ocean is not blue: a visual pigment loss in marine mammals. *European Journal of Neuroscience*, **13**, 1520–1528.

Peitsch, D., Fietz, A., Hertel, H., *et al.* (1992). The spectral input systems of hymenopteran insects and their receptor-based colour vision. *Journal of Comparative Physiology A*, **170**, 23–40.

Pokorny, J. and Smith, V. C. (1982). New observations concerning red-green color defects. *Color Research and Application*, **7**, 159–164.

Pratt, A. E. (1898). *The Use of Color in the Verse of the English Romantic Poets*. Chicago IL: University of Chicago Press.

Rayleigh, L. S. (= Strutt, J. W.) (1890). On defective colour vision. British Association Report, Leeds, pp. 728–729.

Regier, T., Kay, P. and Khetarpal, N. (2007). Color naming reflects optimal partitions of color space. *Proceedings of the National Academy of Science*, **104**, 1436–1441.

Reuter, T. (2011). Fifty years of dark adaptation 1961–2011. *Vision Research*, **51**, 2243–2262.

Rood, O. (1879). *Modern Chromatics, With Applications to Art and Industry*. New York: D. Appleton and Co.

Roorda, A. and Williams, D. R. (1999). The arrangement of the three cone classes in the living human eye. *Nature*, **397**, 520–522.

Sacks, O. (1996). *The Island of the Colorblind*. New York: Alfred A. Knopf.

Sacks, O. and Wasserman, R. (1987). The case of the colorblind painter. *New York Review of Books*, **34**, 25–34.

Schefrin, B. E. and Werner, J. S. (1990). Loci of spectral unique hues throughout the life span. *Journal of the Optical Society of America A*, **7**, 305–311.

Seebeck, A. (1837). Über den bei manchen Personen vorkommenden Mangel an Farbensinn. *Annalen Physik Chemie*, **42**, 177–233.

Sharpe, L. T. and Jägle, H. (2001). I used to be color blind. *Color Research and Application*, **26**, S269–S272.

Sharpe, L. T., Stockman, A., Jägle, H. and Nathans, J. (1999). Opsin genes, cone photopigments, color vision, and color blindness. In Gegenfurtner, K. R. and Sharpe, L. T. (eds), *Color Vision. From Genes to Perception*. Cambridge, UK: Cambridge University Press, pp. 3–51.

Shevell, S. K. and Kingdom, F. (2008). Color in complex scenes. *Annual Review of Psychology*, **59**, 143–166.

Simunovic, M. P., Regan, B. C. and Mollon, J. D. (2001). Is color vision deficiency an advantage under scotopic conditions? *Investigative Ophthalmology and Visual Science*, **42**, 3357–3364.

Smith, V. C. and Pokorny, J. (1977). Large-field trichromacy in protanopes and deuteranopes. *Journal of the Optical Society of America*, **67**, 213–220.

Stabell, U. and Stabell, B. (1982). Color vision in the peripheral retina under photopic conditions. *Vision Research*, **22**, 839–844.

Steward, J. M. and Cole, B. L. (1989). What do color vision defectives say about everyday tasks? *Optometry and Vision Science*, **66**, 288–295.

Stockman, A. and Brainard, D. H. (2009). Color vision mechanisms. In: Bass, M., DeCusatis, C., Enoch, J., *et al.* (eds), *The Optical Society of America Handbook of Optics, 3rd edition, Volume III: Vision and Vision Optics*. New York: McGraw Hill, pp. 11.1–11.104

Stockman, A. and Sharpe, L. T. (2000). The spectral sensitivities of the middle- and long-wavelength-sensitive cones derived from measurements in observers of known genotype. *Vision Research*, **40**, 1711–1737.

Suk, H.-J. and Irtel, H. (2010). Emotional response to color across media. *Color Research and Application*, **35**, 64–77.

Sällström, P. (2001). *Samtal om färgseendets gåta*. Järna, Sweden: Kosmos.

Teller, D. Y. (1998). Spatial and temporal aspects of infant color vision. *Vision Research*, 38, 3275–3282.

Thoen, H. H., Martin, J., How, M. J., Chiou, T.-H. and Marshall, J. (2014). A different form of color vision in mantis shrimp. *Science*, 343, 411–413.

Thompson, D. V. (1956). *The Materials and Techniques of Medieval Painting*. New York: Courier Dover Publications.

Tritsch, M. F. (1993). Color choice behavior in cats and the effect of changes in the color of the illuminant. *Naturwissenschaften*, 80, 287–288.

Valberg, A. (2005). *Light Vision Color*. Chichester, UK: Wiley.

Valdez, P. and Mehrabian, A. (1994). Effects of color on emotions. *Journal of Experimental Psychology: General*, 123, 394–409.

Varela, F. J., Palacios, A. G. and Goldsmith, T. H. (1993). Color vision of birds. In Zeigler, H. P. and Bischof, H. J. (eds), *Vision, Brain and Behavior in Birds*, Cambridge MA: MIT Press, pp. 77–98.

Verhulst, S. and Maes, F. W. (1998). Scotopic vision in colour-blinds. *Vision Research*, 38, 3387–3390.

Verriest, G., Neubauer, O., Marre, M. and Uvijls, A. (1980). New investigations concerning the relationships between congenital colour vision defects and road traffic security. *International Ophthalmology*, 2, 87–99.

Villiers, J. G. d. and Villiers, P. A. d. (1978). *Language Acquisition*. Cambridge, MA: Harvard University Press.

Vingrys, A. J. and Cole, B. L. (1986). Origins of colour vision standards within the transport industry. *Ophthalmic and Physiological Optics*, 6, 369–375.

Wässle, H. (2004). Parallel processing in the mammalian retina. *Nature Reviews Neuroscience*, 5, 1–11.

Wertheim, T. (1894). Über die indirekte Sehschärfe. *Zeitschrift für Psychologie und Physiologie der Sinnesorgane*, 7, 172–187.

Westland, S. (2012). *Frequently Asked Questions about Colour Physics*. Bawtry, UK: Magus Publishing.

Williams, D. R. and Roorda, A. (1999). The trichromatic cone mosaic in the human eye. In Gegenfurtner, K. R. and Sharpe, L. T. (eds), *Color Vision. From Genes to Perception*. Cambridge, UK: Cambridge University Press, pp. 113–122.

Yarbus, A. L. (1967). *Eye Movements and Vision*: New York: Plenum Press.

Zeki, S. (1999). *Inner Vision. An Exploration of Art and the Brain*. Oxford, UK: Oxford University Press.

Index

For EU product safety concerns, contact us at Calle de José Abascal, 56–1°,
28003 Madrid, Spain or eugpsr@cambridge.org.

www.ingramcontent.com/pod-product-compliance
Ingram Content Group UK Ltd.
Pitfield, Milton Keynes, MK11 3LW, UK
UKHW051511240426
470322UK00002B/4